The Potential for Production of 'Hydrocarbon' Fuels from Crops in Australia

by

G.A. Stewart, J.S. Hawker, H.A. Nix,
W.H.M. Rawlins and L.R. Williams

G.A. Stewart - CSIRO Division of Chemical Technology, Canberra

J.S. Hawker - CSIRO Division of Horticultural Research, Adelaide

H.A. Nix - CSIRO Division of Land Use Research, Canberra

W.H.M. Rawlins - CSIRO Division of Chemical Technology, Canberra

L.R. Williams - Chemistry Department, Macquarie University, Sydney

Support for this study was provided under the National Energy Research Development and Demonstration Program which is administered by the Commonwealth Department of National Development and Energy

National Library of Australia Cataloguing-in-Publication Entry

The Potential for production of hydrocarbon fuels from crops in Australia.

Bibliography.

ISBN 0 643 02931 1.

1. Hydrocarbons. 2. Fuel—Australia.
3. Field crops—Australia. I. Stewart, G. A.
(George Alan). II. Commonwealth Scientific and
Industrial Research Organization (Australia).

662′6

 Printed by CSIRO, Melbourne.

81.536-600

TABLE OF CONTENTS

SUMMARY

1. This report is a preliminary assessment of the potential for the production of liquid fuels from resin-containing crops in Australia. None of the potential crops is being grown commercially, and only a small amount of research is being carried out in Australia. More research is being done in USA but no commercial crops are being grown as yet.

2. The lack of commercial experience and field experimental data makes the estimation of crop yields, cost of crop production, and cost of processing more tentative than for previous assessments of alcohols (methanol and ethanol) and seed oils.

3. The candidate crop species, the agro-ecological zones in which they might be grown and predicted yields are shown in Table S1. Cost of production of feedstock was assessed in relation to cost of alternative crops such as wheat and grain sorghum in southern and eastern Australia. For northern Australia, costs were developed for new enterprises. The resin-containing crops could be substituted for existing crops in existing cropping areas, or they could be grown on arable land that is not used for crops at present.

4. It was assumed that costs of recovering resins by solvent extraction would be the same per tonne of feedstock as the cost of solvent extraction of oils from oilseeds. Costs of extraction of resin and rubber from guayule were based on recent US estimates. Likely levels of returns from co-products (lignocellulose, sugars, leaves, rubber) were assessed, and remaining costs were allocated to resins.

5. Excluding the case of guayule, the costs of resins would be $116 to $196 per barrel oil equivalent (see Table S1) and are very high compared with the cost of oil from rapeseed of $75 per barrel o.e. The low proportion of recoverable resin (0.045 to 0.10) means that all crops result in the co-production of large amounts of lignocellulosic residues for which there is no existing market. The most likely use for such large quantities of lignocellulose is conversion to methanol by gasification - catalytic synthesis. This would require ten times as much capital as does resin extraction. However, at current resin contents, the energy content provided by resin would only be a half to a quarter of methanol energy, i.e. resin would be a by-product. Australia has large amounts of existing lignocellulosic residues that could be converted to methanol, so there is no point in developing new crops primarily as sources of lignocellulose for conversion to methanol. If resin content could be increased to 0.15 to 0.20, the energy content due to resin would equal the energy content due to methanol. However the residue of a medium sized plant for processing resin-containing crops would be only one fourth of that needed for a minimal methanol plant. We conclude that returns from resin-containing crops should be based on sale of resin only. Without revenue from the sale of bagasse, resin costs from candidate species in Table S1, excluding guayule, would be $151 to $270 per barrel o.e.

Table S1 Candidate crop species for various regions in Australia, and their estimated yields and costs

Candidate crop species	Region rainfall season	Estimated yield t(DW)/ha per year	Estimated feedstock cost $/t(DW)	Proportion recoverable resin° in DW	Cost of resin $/barrel o.e.***
Euphorbia lathris	southern, winter	3	57	.07	147
Euphorbia lathyris	southern, irrigated	10	51	.07	134
Asclepias rotundiflolia	eastern year round	7.5	58	.09	116
Calotropis procera	northern, summer	10	46	.045	190
Calotropis procera	northern, irrigated	20	48	.045	196
Parthenium argentatum	Tara Shire	5	60*	.10	-79¶
(guayule)			184**	.10	104

* Cost estimates of this report
** Cost estimates for Texas, USA
*** barrel o.e. = barrel oil equivalent = 5.8GJ. Resins assumed 40GJ/t = 6.9 barrels oil equivalent per tonne.
o recoverable resin includes resin I and resin II for Euphorbia lathyris and Parthenium argentatum, and resin I plus hydrocarbons for other species.
¶ Revenue from other products exceed total costs by $79/barrel o.e.

6. For the technologies considered in this report, and without sales of bagasse, the recoverable resin content of these types of crops would need to be increased to 0.15 to 0.20 of dry matter to become competitive with seed oils (see Fig. 8-2). Several Eucalyptus spp. have essential oils plus resin contents greater than 0.12 of dry matter, and are worthy of further investigation.

7. Two areas of research may lead to higher resin contents:
(a) crops harvested near maturity appear to have appreciably higher resin contents, and short-term field experiments are needed to confirm this (see 8 below); and

(b) further chemical research is needed to ascertain if further components of the plants (acetone soluble-heptane insoluble) have high energy contents and whether they could be converted catalytically to liquid fuels. Appropriate analytical methods should then be identified for application to existing plant samples and those that arise from (a) above.

8. Crop yields have been estimated indirectly, and limited small scale field experiments at two or three sites are essential to provide some direct information.

9. In the USA some thousands of species have been examined for resin and rubber content, and only two or three species have been identified with high resin content. Plant physiology - biochemistry does not provide any good leads on likely candidate species. The use of ether extract analysis as

performed for animal feedstuffs appears to offer promise as a survey tool because of the very large number of species that have been analysed by the method. At the same time it appears desirable to have a small continuing program of collection of possible candidate species by biologists who are adequately informed on key aspects of sampling.

10. The production of resin from guayule appears very favourable, primarily because guayule produces twice as much rubber as resin and the rubber is about five times as valuable as crude oil. An ANU research fellow funded by the CSIRO Division of Land Use Research has visited the US and Mexico to gather further information on guayule. A more comprehensive assessment of guayule will be made later this year, which will take into account existing interests and activities of the State Departments of Agriculture.

11. The resins are dark coloured solids that melt above ambient temperature into viscous liquids. They have not been tested directly as fuels for diesel engines. They can be catalytically cracked into conventional liquid fuels (gasoline, LPG, and distillate), but the cost of that conversion is not known. *Eucalyptus* spp. also contain essential oils that have high octane ratings and are premium fuels for spark ignition engines.

1. INTRODUCTION

The possibility of producing crops containing 'hydrocarbons' that could be extracted for use as liquid fuels was first identified by Calvin (1974), and subsequently received wide publicity around the world. At present the only hydrocarbon-producing crop in world agriculture is Hevea brasiliensis, the source of natural rubber. That product is very valuable in its existing uses, being five times higher in cost than crude oil. Guayule (Parthenium argentatum), a perennial shrub native to northern Mexico and southwestern USA, was exploited as an alternative source of rubber in the period 1903 to 1950, largely by harvesting wild stands of the shrub in Mexico. It also contains appreciable quantities of resins that could be used as sources of liquid fuels. Most of the recent research on resin-containing crops in the USA has been on Euphorbia lathyris (gopher plant).

Of hundreds of species of plants that have been surveyed for 'hydrocarbons' or 'resins', only a small number of species have been identified as potential crop species. None are cultivated and considerable research and development will be needed to bring them to the stage of commercial development. Very little information is currently available on their production and processing. Thus this assessment is necessarily more generalised and less precise than the earlier assessments of potential production and costs of alcohols from agriculture and forestry (Stewart et al. 1979a,b) and of seed oils as diesel fuels (Stewart et al. 1981), which were based on existing technologies of crop production and processing.

Only a small proportion of extracted materials are true hydrocarbons (i.e. contain only hydrogen and carbon). The dominant components are most commonly polycyclic terpenoids that resemble resins from wood and so the term 'resin' is mostly used in this report (see Section 2). The crude resins from 200 species of plants were described as dark solids that melt slightly above room temperature to form viscous liquids (Buchanan et al. 1978c).

This report:

. reviews laboratory extraction procedures and the nature of the materials extracted with emphasis on the gopher plant and guayule;

. reviews investigations of resin crops in Australia and identifies candidate crop species;

. considers the land suitable for various crops, appropriate farming systems, and arrives at a cost of production;

. estimates the cost of solvent extraction of resins, and the cost of resins in relation to seed oils; and

. identifies major weaknesses in the data and the needs for research.

2 METHODS OF LABORATORY EXTRACTION AND NATURE OF RESINS

2.1 Methods of Laboratory Extraction

Resins, hydrocarbons, and other components of plant matter are separated from dried, finely ground biomass by various organic solvents in Soxhlet refluxing extractors. The nature and amount of material extracted is determined by the nature of the solvent used, or the sequence in which a number of different solvents are used. Further separation is commonly carried out by mixing the extractives with two immiscible solvents. Various components distribute themselves between the two solvents according to their distribution coefficient - the ratio of their concentration in the two solvents which reflects their preferred solubility in one solvent over the other.

The methods of partitioning that have been used for resins and hydrocarbons are shown in Fig. 2-1. The following terms are used for various fractions in this report:

Resin I = acetone soluble - heptane soluble, higher heating value (HHV) = 40 to 42MJ/kg

Resin II = acetone soluble - heptane insoluble (*Euphorbia lathyris* and *Eucalyptus* spp.), HHV = 31 to 33MJ/kg

'Polyphenols' = acetone soluble - heptane insoluble (other species), HHV = 22MJ/kg

Hydrocarbons = heptane soluble - acetone insoluble, HHV ≧ 40MJ/kg

Essential oils = that part of Resin I recoverable by steam distillation, HHV = 42MJ/kg

Benzene and cyclohexane have been used as alternatives to heptane in extraction of true hydrocarbons from acetone extracted residues, or extraction of resin I plus true hydrocarbons from dried plant material. Hexane has been used as an alternative to heptane in separating resin I from other acetone extractables.

2.1.1 Edison's method of hydrocarbon and resin extraction

Calvin (1978) reported that Thomas A. Edison, in searching for alternative plant sources of rubber, analysed thousands of samples by acetone extraction followed by benzene extraction (see Fig. 2-1a). The acetone removed oxygen-containing compounds which were termed oil (resin I + resin II or polyphenols in our terminology), and the benzene dissolved the true hydrocarbons including natural rubber, for which the molecular weights were determined.

FIGURE 2.1

Some schemes used in laboratory solvent extraction of hydrocarbons, resins and other fractions. Our standard fraction names are shown in capitals. Oil fraction in (b) equals our resin 1.

(a) Edison, from Calvin (1978)
(b) Buchanan *et al.* (1978c)
(c) Nemethy *et al.* (1979)
(d) Nemethy *et al.* (1980)
(e) Peoples and Johnson (1980)
(f) Nishimura and Calvin (1979)

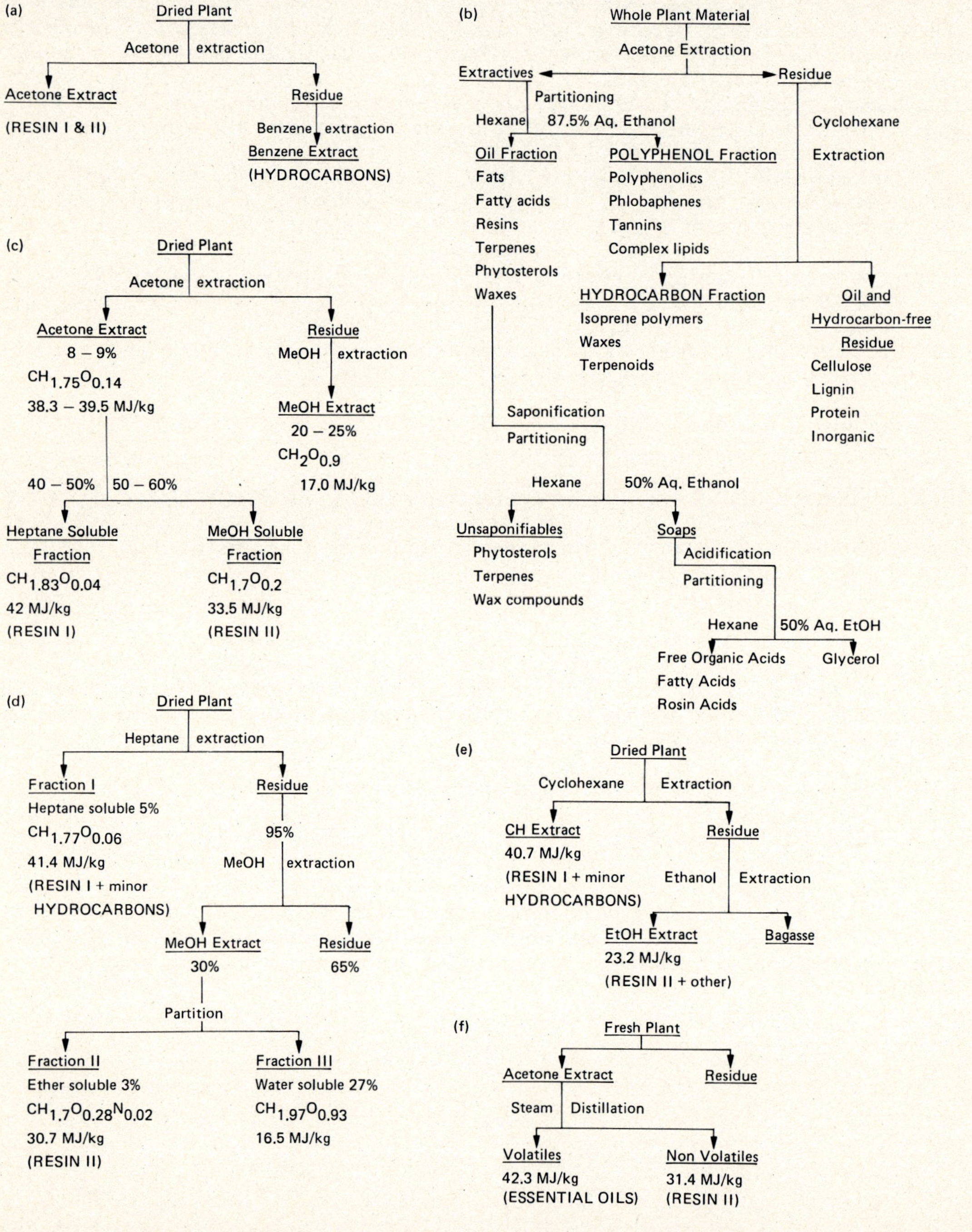

2.1.2 Method of plant extraction and partitioning used by USDA, Peoria

In 1974 the Northern Regional Laboratory of the US Department of Agriculture at Peoria, Illinois, instituted a screening program to identify and evaluate species as sources of 'plant oils' (resin I in our terminology), hydrocarbons, protein and fibre. The results of their analysis were incorporated into a simple numerical rating system which placed most emphases on botanical evaluation for cultivation and content of resin and hydrocarbons.

Their method of partitioning (see Fig. 2-1b) varied slightly over time to avoid the more toxic solvents, but these changes were reported to have no effect in analytical values (Buchanan et al. 1978b). The dried plant material was extracted first with acetone. The residue was then extracted with cyclohexane (benzene:hexane 2:3 was used by Buchanan et al. 1978a) to remove the true hydrocarbons, which were either natural rubber and/or waxes in those fractions that were further examined by infrared spectroscopy (Buchanan et al. 1980c).

The acetone extract was partitioned between hexane and aqueous ethanol (aqueous methanol in Buchanan et al. 1978a). This separated the less polar resin I (plant oil) fraction from the more polar 'polyphenols'. The resins were further partitioned into various components as shown, and the results of that partitioning will be described in Section 2.5. Crude protein content and fibre utility were determined separately. The extraction method did not give an estimate of the content of sugar, nor was a separate analysis made.

2.1.3 Methods of plant partitioning used at the University of California, Berkeley

Calvin's laboratory at Berkeley, California initially used acetone followed by benzene extraction (i.e. as Edison) for a survey of hydrocarbon-like material in 12 species (Nielson et al. 1977). The acetone fraction was not partitioned with immiscible solvents, but the amount of major compounds (glycerides, isoprenoids and terpenoids) were estimated by nuclear magnetic resonance spectra. In subsequent studies of *Euphorbia lathyris* (Nemethy et al. 1979, 1980) the extraction with benzene for true hydrocarbons was not used (see Fig. 2-1c and d), and that appears justifiable as the content of true hydrocarbons in the whole plant was 0.37% (Buchanan et al 1978c) and of leaf was 0.3% (Nielson et al. 1977). Two different partitioning schemes were used, but each ended up with three fractions that were virtually identical in proportions of C,H and O. In one case (Nemethy et al. 1979) the dried material was first extracted with acetone, and the extract was partitioned into heptane soluble and methanol soluble fractions which we term resin I and resin II. The residue from acetone extraction was extracted with methanol and gave a low energy material with an elemental composition close to C_6 sugars. In the second case (Nemethy et al. 1980) the first extraction was with heptane and that resin I plus minor hydrocarbons was very similar in composition to the resin I of the previous procedure. The plant residue from heptane extraction was extracted with methanol and the methanol extract was then partitioned with ether and water. The ether fraction II was very similar to the resin II of the previous system and the water soluble fraction III was very similar to the methanol fraction of the previous system.

2.1.4 Resin extraction by the Arizona Method

In 1978 the Diamond Shamrock Corporation provided funds to the Office of Arid Land Studies at the University of Arizona to examine the possibility of producing energy raw materials or chemicals from arid zone plants (Johnson and Hinman 1980). Their system for partitioning of E. lathyris (see Fig. 2-1e) was similar to the first part of Nemethy et al (1980) but used cyclohexane and ethanol rather than heptane and methanol (Peoples and Johnson 1980). The energy content of the cyclohexane extract was similar to resin I of Nemethy et al. (1979, 1980). The ethanol extract has an energy content between that of fractions II and III of Nemethy et al. (1980). Peoples et al (1981) appreciated that this was a mixed fraction as they reported that E. lathyris produced an energy-rich, hydrocarbon-like substance and additional lower energy material that excluded carbohydrates, but they did not report how the later component was assessed.

2.1.5 Extraction of Essential Oils and Resins from Eucalyptus spp.

Nishimura and Calvin (1979) used fresh plant material in the extraction of essential oils and resins from various parts of Eucalyptus globulus by exhaustive (two days) extraction with acetone in Soxhlet extractors. The essential oils were obtained by steam distillation of the acetone fraction (see Fig. 2-1f), and a variety of sophisticated chemical techniques were used to identify the mono- and sesqui-terpenes of the essential oils. The other acetone extractables were not partitioned in any way. The same techniques were used to examine four other species of Eucalyptus spp. by Nishimura et al. (1980).

2.2 Chemistry and Biosynthesis of Resins and Rubber

The following notes from Roberts and Caserio (1965) provide for nonchemists a simple description of the nature, terminology, and biosynthesis of resins and rubber.

Resins and rubber are both isoprenoid compounds, i.e. they are made up of units of isoprene C_5H_8 (See Fig. 2-2). Isoprene does not occur naturally, but only as parts of larger compounds.

Terpenes are hydrocarbons with multiples of C_5H_8, and a double bond is lost for each junction between isoprene units (see myrcene in Fig. 2-2). The following terms are used for those multiples:

Terpenes	C_{10}
Sesquiterpenes	C_{15}
Diterpenes	C_{20}
Triterpenes	C_{30}
Tetraterpenes	C_{40}

The terpenes may be either open chains, as is myrcene, or cyclic as in the bicyclic terpene, α pinene (see Fig. 2-2), which is the main constituent of oil of terpentine. The terpenes and sesquiterpenes are known as essential oils.

FIGURE 2.2

Structure of isoprene and some examples of terpenes, terpenoids and sterols

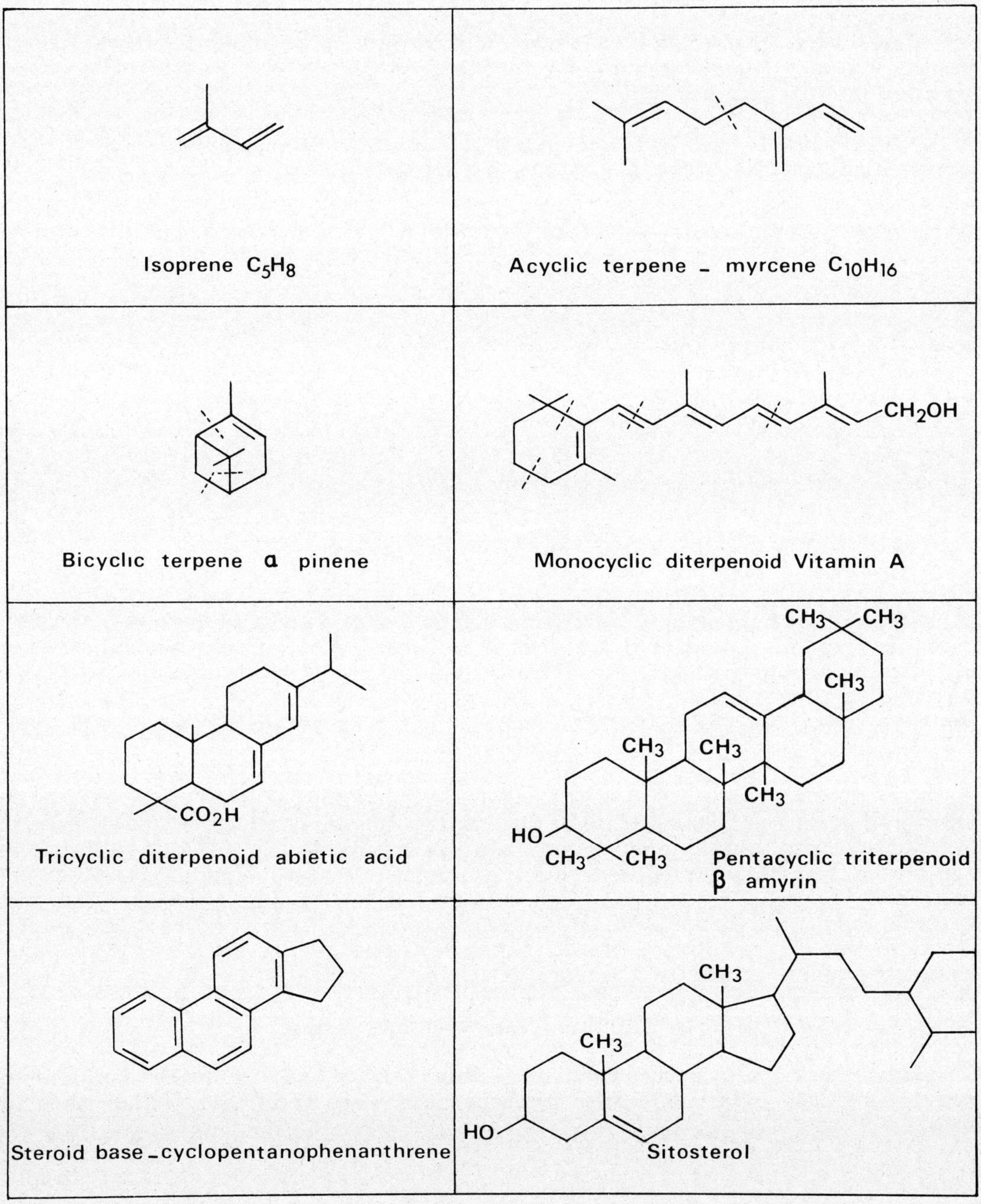

Natural rubber from Hevea and guayule is an amorphous polymer of isoprene $(C_5H_8)_n$, and n is commonly about 20 000.

Terpenoids are compounds that contain some oxygen, either as alcohols (see vitamin A in Fig. 2-2), aldehydes, ketones, or carboxylic acids. Abietic acid (Fig. 2-2) is a tricyclic terpenoid and is the main component of rosin from pine trees. β amyrin, a common resin in plants, is pentacyclic.

Sterols are alcohols of the terpenoid group of compounds which have a common base of cyclopentanophenanthrene (see Fig. 2-2). Sitosterol is one of the many sterols found in plants.

Acetic acid is the common starting point for the biosynthesis of both terpenes and sterols. The main steps are set out below:

$$\text{acetic acid + coenzyme A} \xrightarrow{\text{ATP}} \text{acetyl CoA}$$

$$\text{acetyl CoA} + CO_2 \longrightarrow \text{malonyl CoA}$$

$$\text{acetyl CoA + malonyl CoA} \xrightarrow[\text{-COA}]{-CO_2} \text{acetoacetyl CoA}$$

$$\text{acetoacetyl CoA + acetyl CoA} \longrightarrow \text{mevalonic acid}$$

$$\text{mevalonic acid} \xrightarrow[-CO_2\ -\ H_2O]{\text{ATP}} \text{isopentyl phosphate}$$

$$\text{2 x isopentyl phosphate} \longrightarrow \text{geranyl pyrophosphate}$$

Geranyl pyrophosphate contains 10 carbons. If hydrolysed, it would give a terpene. Repeated reactions between geranyl pyrophosphate and isopentyl phosphate adds 5 carbons to the chain to build up to higher terpenes or to sterols.

2.3 Nature of Resins in Euphorbia lathyris

Nemethy et al. (1979) examined the composition of the resin I fraction extracted from Euphorbia lathyris directly by heptane, which was very similar to the heptane soluble portion of the acetone extract (see Fig. 2-1). The heptane extract was first separated by column chromotography on silica gel into five fractions by elution with solvents of increasing polarity. Infrared spectra were used to determine functions groups in each fraction. Except for fraction 1, all fractions were either further separated by thin layer chromotography (TLC), saponified, or derivatised prior to gas chromotography - mass spectroscopy analysis. High resolution mass spectroscopy was used to determine elemental composition.

The results are summarised in Table 2-1. In all, 45 major components were recognised. The only true hydrocarbons were fraction 1, which made up only 0.07 of the extract. It consisted of C_{31} and C_{33} paraffins. Apart from a small amount of fatty alcohols in the sixth subfraction of fraction 3, all the remainder of the extract was composed of polycyclic triterpenoids (i.e. C = 30) functionalised as fatty acid esters, ketones or alcohols, and sterols. The whole plant of E. lathyris contained a much greater variety of triterpenoids than the latex. Only about 10 to 15% of the total amount of resins occurred in the latex (E. Nemethy: Personal communication).

Resin II, the fraction of the acetone extract that was not soluble in heptane but was soluble in methanol, had an elemental analysis of $CH_{1.7}\,O_{0.2}$ (see Fig. 2-1e). Similar material was obtained as ether solubles from the methanol extract of material insoluble in heptane as the first solvent, with an elemental analysis of $CH_{1.7}O_{0.28}$ (see Fig. 2-1d). That more highly oxygenated fraction was not analysed in detail, but consisted mostly of bifunctional triterpenoids (E. Nemethy: personal communication). This fraction is called resin II because its elemental composition is very similar to that of fraction 5 in resin I (see Table 2-1).

The portion of the methanol extract not soluble in ether consisted mainly of sugars, with some amino acids and some unknowns including flavanoids. The elemental analysis of $CH_2\,O_{0.9}$ is close to that expected from such a mixture,

Table 2-1 Proportion, elemental composition, number of major components and general character of the five main fractions of the heptane extract (resin I) in dry matter of *Euphorbia lathyris* (Nemethy *et al*. 1979)

Eluent	Proportion of total	Elemental composition	Number of major components	Class of compounds
1 Heptane	.07	CH_2	2	Hydrocarbon
2 Benzene	.33	$CH_{1.72}O_{0.03}$	5	Fatty acid esters of triperpenoids
3 Ethyl acetate	.41	$CH_{1.67}\,O_{0.07}$	25	Tetra and pentacyclic triterpenoids, ketones, alcohols
4 Acetone	.05	$CH_{1.63}\,O_{0.15}$	8	Phytosterols and bifunctional compounds
5 Methanol	.14	$CH_{1.69}\,O_{0.22}$	5	Bifunctional triterpenoids

2.4 Nature of Resins in Guayule (*Parthenium argentatum*)

The components of guayule resins are illustrated in Fig. 2-3. Resins extracted by acetone at the Saltillo pilot plant (Campos-Lopez *et al*. 1978) are of two kinds. One is from direct deresination of the stem, branches and roots, and the other is from the deresination of the rubber crumbs after rubber has been separated from the bagasse by flotation. The former contained 3% essential oils and 6% saponifiables whereas the latter contained 5 and 17% respectively (Belmares-Sarabia and Jimenez-Valdes 1978).

Figure 2-3

Components of guayule resin (National Academy of Sciences 1977)

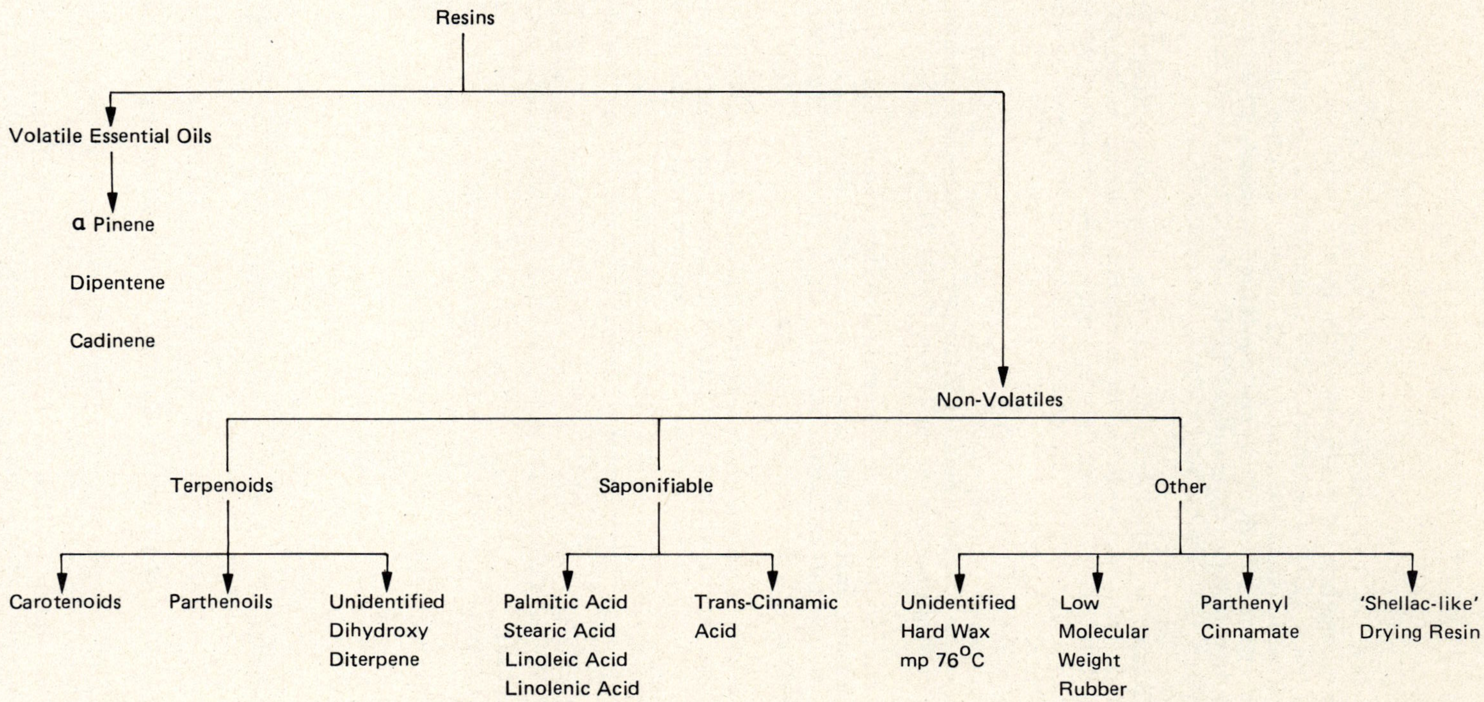

The saponifiables contained 56% linoleic acid and 10 to 12% of palmitic, oleic and linolenic acids as triglycerides and 10% of transcinnamic acid (Campos-Lopez et al 1978). A 'drying resin' constitutes 35 to 50% of shrub resin, but a much smaller proportion of that resin is from rubber crumbs (Hammond and Polhamus 1965). It is hexane-insoluble, alcohol-soluble, shellac-like gum resin which can be easily polymerised to a heat-resistant clear coating of good solvent resistance. No data is available on the proportion of other constituents. Research workers in Mexico are examining the possibility of utilising the resins for varnishes and adhesives, and other possible uses such as pigment dispersors and tackifiers for rubber.

Buchanan et al. (1978c) reported the analysis of one year old guayule from California which presumably included leaves (see Table 2-2). They concluded that the resin I fraction contained appreciable amounts of essential oils because the sum of saponifiables plus unsaponifiables (see Fig. 2-1a) was only 0.76 of total resin I, compared with over 0.9 for most species analysed. The remaining material was mainly nonglyceride esters (probably terpenoids), hydrocarbons and triglycerides.

2.5 Nature of Resins in Other Plant Species

Buchanan et al. (1978c) carried out less detailed studies of the nature of heptane soluble part of the acetone extract (= resin I) from fourteen plant species that appeared promising for the production of resins and rubber, with 5 to 10% of polymeric hydrocarbons. Their scheme for partitioning of biomass, a modification of the classic sequential acetone-benzene extraction system, is illustrated in Fig. 2-1 and their results are presented in Table 2-2. The green pigments and polar materials were removed from the plant oils. Recovery was 80 to 90% and refined resins were clear or straw to orange coloured. The composition of unsaponifiable matter was estimated from TLC and densitometry. The yield of unsaponifiable matter plus free fatty acids was normally more than 0.90 of the resin, except for guayule (Parthenium argentatum), which contains significant quantities of volatile essential oils (Campos-Lopez et al. 1978). The unsaponifiable matter from most whole plant extracts contained only 3 TLC fractions corresponding to sterols, less polar free alcohols and hydrocarbons. Ten of the fourteen resins had high contents of nonglyceride esters, less polar than triglycerides, and these probably corresponded to triterpenoids and wax esters. Three resins were dominated by triglycerides, and the authors commented that, in at least one case, this would have been due to a relatively high proportion of seed in the sample. One species, Sassafras, was dominated by free alcohols that were considerably less polar than characteristic wax alcohols.

Summarising the composition of resin I from vegetative material:

. volatile essential oils were low in all except guayule and Eucalyptus spp.

. hydrocarbon content was generally low, and probably included some terpenes as in guayule and some low molecular weight rubbers soluble in acetone

. triglycerides rarely dominated

. nonglyceride esters, probably polyterpenoids or wax esters, mostly dominated.

Table 2-2 Lipid classes in heptane solubles of acetone extract of 12 species from Illinois and 2 species (*Parthenium argentatum* and *Euphorbia lathyris*) from California (Buchanan *et al*. 1978c)

Plant Source	Harvest date	Oil content of DW %	Oil composition					
			Sterols %	Other free alcohols %	Free acids %	Triglycerides %	Nonglyceride esters %	Hydrocarbons %
Ambrosia trifida	9/20/77	4.21	11	4	7	68	7	3
Asclepias incarnata	7/28/77	2.70	9	15	8	13	48	7
Asclepias syriaca	7/27/77	4.46	5	11	5	trace	72	7
Cacalia atriplicifolia	8/3/77	2.99	10	24	7	10	43	6
Campanula americana	9/8/77	6.51	13	5	10	58	10	4
Cirsum discolor	10/5/77	5.66	2	4	4	21	67	2
Eupatorium altissima	9/20/77	6.08	6	5	7	36	44	2
Euphorbia dentata	8/25/77	4.13	6	6	5	42	36	5
Euphorbia lathyris	3/25/77	9.21	3	20	18	3	49	7
Parthenium argentatum (guayule)	5/13/77	4.04	10	7	5	23	31	24
Phus glabra	8/18/77	5.10	12	11	19	13	39	6
Sassafras albidium	9/8/77	2.26	8	46	--	2	28	16
Sonchus arvensis	6/10/77	4.63	5	19	5	5	60	6
Vernonia altissima	5/23/77	2.62	13	12	2	4	68	1

Buchanan et al. (1978a,b,c) used the term 'polyphenol fraction' for the acetone extract that was not soluble in hexane (see Fig. 2-1b), i.e. as resin II of Nemethy et al. (1980). They suggested that it contained polyphenols, phlobaphenes, tannins, and complex lipids. The only elemental composition reported was for three samples, the range being 52.1 to 57.0% for carbon and 6.7 to 7.7% for hydrogen (Buchanan et al 1978a). Taking the mean of those ranges and assuming the remainder was all oxygen, the elemental composition would be $CH_{1.59}\ O_{0.53}$. This contains much more hydrogen than the reported standard compound, quercetin, $CH_{0.67}\ O_{0.47}$. It also contained appreciably more oxygen than resin II from E. lathyris. The species for which samples were analysed were not listed. E. lathyris was not amongst the 16 species for which other analytical data were given. In view of the lower calculated energy content (22MJ/kg) and the uncertainty of the constituents of this fraction, the term "resin II" for this fraction will be used only for E. lathyris, and for all other species the term 'polyphenol fraction' will be used. Some 12 species contain over 0.10 of 'polyphenols'. Further analysis is needed to determine the constituents of this fraction, their energy content, and the feasibility of utilising them as high energy fuels.

2.6 Nature of Essential Oils and Resins in Eucalyptus spp.

The essential oils of E. globulus are dominated by sesquiterpenes in the fruits and monoterpenes in the leaves, branches and shoots (Nishimura and Calvin 1979). The essential oils of Eucalyptus spp. are mostly dominated by cineole, but some species or variants are dominated by piperitone and/or phellandrene, and a few species are dominated by perfumery oils such as citronellal (Small 1977). No data is available on the chemical nature of the non-volatiles, but their energy content is 31.4MJ/kg (Nishimura et al. 1980), i.e. similar to the resin II fraction of Nemethy et al. (1980).

2.7 Possible Use of Ether Extraction for Resins

Nemethy et al. (1980) reported that acetone tended to dissolve some carbohydrates in refluxing Soxhlet extractors, and implied that it led to variable results from acetone extraction. If the resins extractable in heptane were also extractable in ether, the ether solvent extraction as widely used in animal feeds would give a simple laboratory method of extracting all resins. As well it would facilitate searching a very large number of analyses of animal feeds for crop materials with high resin or seed oil content. One of us (L.R.W.) did a heptane extraction on Calotropis procera before and after extraction with ether and found that ether extracted slightly more material than hexane, and that hexane also extracted a small amount that was not extracted by ether, possibly true hydrocarbons. Table 2-3 shows that three US browse shrubs, not previously mentioned as sources of resins, have an ether extract of about 0.10 of DW. A search of Australian animal feed data is now being initiated.

Table 2-3 Proportion of ether extract in dry weight for three US browse shrubs from animal feed analysis (National Academy of Sciences 1969)

Botanical name	Common name	Ether extract (proportion of DW)
Artemisia tridentata	Big sagebrush	.101
Salvia mellifera	Black sage	.094
Chrysothannus stenophyllus	Yellow brush	.122

2.8 The Role of Resins and Rubber in Plant Physiology

The classical hydrocarbon plant, Hevea brasiliensis, produces a latex that can be tapped from lactifers. The dry matter of the latex is 87% rubber, but Euphorbia spp. and Asclepias spp. had only 1 to 12% rubber and 30 to 75% sterols in their latex dry matter (Nielsen et al. 1977).

Only about 10 to 15% of the resins of E. lathyris is contained in the latex (E. Nemethy: personal communication) and the number of triterpenoids is much smaller than in the whole plant (Nemethy et al. 1979). The content of resins in stems was only about 0.03 and was much lower than the level in the leaves, which was about 0.08 (Sachs et al. 1981). They found that irrigated E. lathyris contained only 0.044 of resins compared with 0.062 in a dryland crop. However this could be explained entirely by the higher proportion of leaf in the dryland crop, and there was no evidence of stimulation of resin formation by water stress, as had been reported for rubber formation in guayule.

Guayule can contain up to 0.26 of rubber, but it is within soft-walled discrete cells in the stems and roots and cannot be recovered by tapping. Traub (1946) and Benedict (1949) reported that the rubber could not be remetabolised when plants were under severe energy stress. The formation of rubber is stimulated by moisture or low temperature stress (Bonner 1943). In moist equable climates, e.g. Southern Texas, guayule grows relatively rapidly but has low rubber content (Hammond and Polhamus 1965). Guayule bark has 0.07 to 0.1 of its dry weight as resins, but wood and roots have lower contents. Leaves have about 0.1 of the dry weight as resins (Curtis 1947). Benedict et al. (1947) found a seasonal variation in resin content that was not affected greatly by moisture stress.

Application of bio-inductors can increase the production of latex and rubber in Hevea (Dickenson 1976) and can increase the content of rubber in guayule (Yokoyama et al. 1977), but we are not aware of any reports of increased resin content or yield.

Asclepias rotundifolia has approximately equal contents of resins in leaf and in stem.

There is no definite evidence of the physiological or biochemical significance or role of rubbers and resins in plants, so those aspects cannot be used as a guide to the identification of promising crop species.

2.9 Conversion of Resins and Rubber to Liquid Fuels

Weisz et al. (1979) reported that Mobil's zeolite catalyst ZSM5 could be used for converting methanol, various vegetable oils, Hevea brasiliensis latex, and dipentene to mainly liquid fuel components, mostly C_3 and C_4, and C_6 to C_9. The latter were mostly benzene, toluene and xylene, which gave a gasoline with a research octane number of 90 to 96. 'The selectivity results from the fact that the terminal size of the products is limited by the molecular shape selectivity of the catalyst, that is, by the constraining size of the intracrystalline cavities'. The material was fed in a hydrogen gas stream over the catalyst in a reactor held at 400 to 500°C. When the Hevea latex was fed into an empty reactor at 400°C it broke down into C_5, C_{10}, C_{15} and C_{20} units, When Hevea latex was fed directly to the catalytic reactor 26% of the rubber was reduced to coke. When a C_{10} compound, dipentene, was fed directly it gave a product mix similar to that obtained by processing methanol. The C_5H_8 unit, isoprene, is the basic building block for both rubber and terpenoids, so it is not surprising that rubber and resins break down to C_5 (isoprene) and small multiples of it. The heptane soluble resins from Euphorbia processed with the Mobil catalyst (Nemethy et al. 1980) gave a product mix with peaks in the C_{2-3} and C_{7-8} ranges (see Fig. 2-4). The resin II fraction (ether soluble) is similar in elemental composition to the more highly oxygenated material in resin I, and it would be reasonable to assume that they could be similarly converted.

FIGURE 2-4

Products of conversion over Mobil zeolite catalyst of Euphorbia lathyris terpenoids (heptane soluble) (adapted from Nemethy et al 1980).

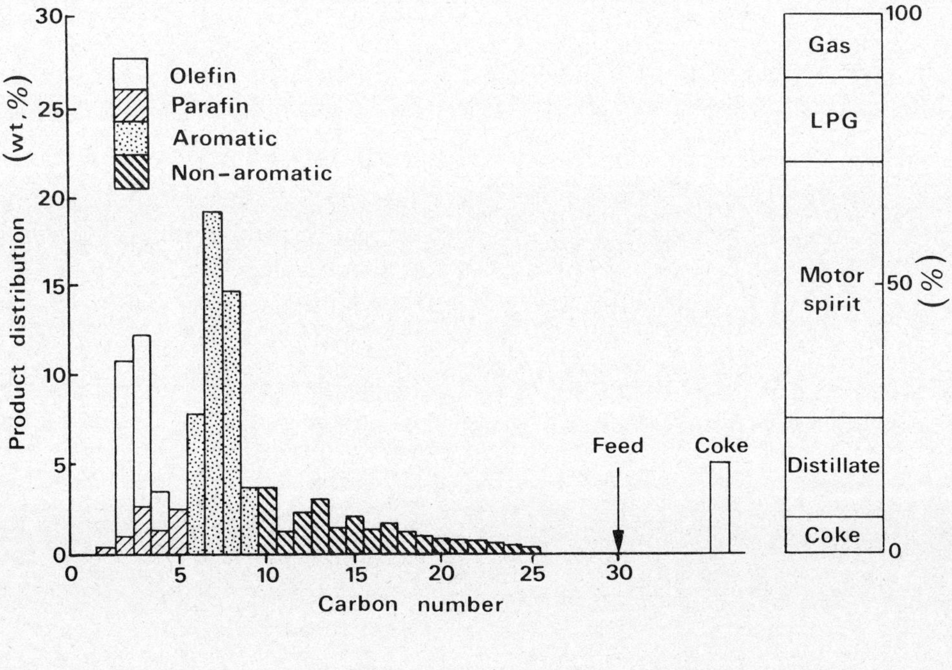

Weisz et al. (1979) reported that the efficiencies in conversion of oil and coal derived materials to liquid fuels increased for compounds with higher hydrogen-carbon ratios and lower heteroatom (e.g. oxygen, sulphur) content. Efficiency was not defined, but it probably is related to the proportion of the feedstock energy recovered in the desired gaseous and liquid fuel products and possibly the amount of hydrogen used up. Formation of coke would be indicative of low efficiency.

The polyphenol fraction (Buchanan et al. 1978a,b,c) has lower hydrogen and higher oxygen content than resin II, and would thus give lower efficiency of conversion. The nature of the compounds in that fraction is not known, nor whether those compounds, or first stage pyrolysis products from those compounds, would be sufficiently stable and small to be processed by this catalyst.

2.10 Essential Oils from *Eucalyptus* spp. as Liquid Fuels

The essential oil of *Eucalyptus* spp. was tested in a small spark ignition engine (Takeda 1979). Performance was very good except that there were problems starting a cold engine on straight *Eucalyptus* oil. These could be readily overcome by adding 20 to 30% of alcohol or gasoline. Research octane rating of *Eucalyptus* oil was 100. Viscosity was higher than for gasoline, but an oversize main jet nozzle in the carburetor overcame that problem. *Eucalyptus* oil has a lower calorific value per unit weight than gasoline (41.2 vs 46.5 MJ/kg), but its higher specific gravity (0.9166 vs 0.735) results in higher calorific values per unit volume (37.8 vs 34.2 MJ/litre). Engine thermal efficiency was slightly high for *Eucalyptus* oil (mean 14.9 MJ/HP compared with 16.1 MJ/HP for gasoline), but this may have been due to better engine tuning as the gasoline engine gave 4 to 6% carbon monoxide compared with around 1% for *Eucalyptus* oil. This type of *Eucalyptus* oil appears to be a premium quality finished fuel for a spark ignition engine.

3 SURVEY OF RESIN-CONTAINING SPECIES

3.1 Content of Resins and Similar Compounds in Euphorbia lathyris

The known analytical data for E. lathyris are shown in Table 3-1. The data for Sachs et al. (1981) are based on a large number of samples taken apparently at a single harvest time when the crop was still vegetative. The extreme range of values for whole plant resin I fraction was 0.032 to 0.082, and the averages for irrigated and unirrigated crops were 0.044 and 0.062 respectively. However that difference was completely explainable by the difference in proportion of leaf and stem and the relatively high content of resin in the leaves.

Whole plant material was used by Nemethy et al. (1979, 1980) and mention was made of the different content of resins in leaf and stem, so probably the plant material was vegetative. The number of samples analysed was not stated. The stated resin I content of 0.05 is similar to that of Sachs et al. (1981), and it is associated with 0.03 of resin II, giving a total content of acetone extractables of 0.08.

The acetone recovery of the single sample analysed by Buchanan et al. (1978c) is roughly twice that of Nemethy et al. (1980) and the difference is approximately equivalent in energy content to that of the sugars reported by Nemethy et al. (1980). The sample analysed by Buchanan et al. (1978c) was provided by Professor M. Calvin and was reportedly harvested on March 25. Sachs et al. (1981) sowed E. lathyris in April and 19 days were required for emergence. Thus a plant harvested in March must have overwintered. If grown in an unheated glasshouse or in the open in central California the low winter temperatures would have then provided the chilling necessary to induce flowering and seed production. However Buchanan et al. (1978c) found very low levels of triglycerides and this would suggest that the formation of seeds had not commenced.

The proportion of acetone extract in leaves reported by Nielson et al. (1977) appears relatively high. This sample was from the herbarium collection of the Botany Department of the University of California, Davis. Herbarium specimens normally contain stem, leaves and flowers and/or seed. That an analysis was made for seed indicates that seed was present, and the high content of triglycerides would indicate that the seeds were nearly mature. If the plants from which the herbarium specimen was taken had 15% seed, 30% leaf and 55% stem, the overall plant content of acetone extractables would have been 0.16 of which about 70% were reported to be triglycerides. However, while the total is consistent with that of Buchanan et al. (1978c), the high content of triglycerides is not, and there are no data that indicate that terpenoids can be converted to triglycerides. The only overall conclusion that can be drawn from the data for these two samples is that resin content appears to be much higher in plants approaching physiological maturity.

E. lathyris has occasionally been grown as an oilseed crop in Russia, China and Japan, the seed capsules do not dehisce explosively at maturity as do other Euphorbia spp., and senescent leaf material is apparently not shed (Hinman et al. 1980). Thus it appears very desirable to test whether E. lathyris grown to near physiological maturity (i.e. almost as an oilseed crop) will have total resin content in the range of 0.15 to 0.20, which would probably allow it to compete with oilseeds (see Fig. 8-2).

Table 3-1 Analytical data for *Euphorbia lathyris* in USA

	Number of analyses	Proportion of Dry Weight: Acetone extract	Resin I (or plant oil)	Resin II (or polyphenol)	Plant Part	Comment
Nielsen *et al*. 1977	1	.25			leaves	if whole plant consisted of 15% seed, 30% leaves, 55% stem the average proportion of acetone extractables would be 0.16
" "	1	.40			seeds	
" "	1	.045			stem	
Buchanan *et al*. 1978c	1	.163	.092	.071	whole plant	harvested M. Calvin Mar. 25, probably near flowering
Nemethy *et al*. 1979	Several	.08-.09			whole plant	
" " 1980	"	.08	.05	.03	" "	the whole plant contained 0.20 of sugars which is equivalent in energy to 0.08 resins
Sachs *et al*. 1981	n.a.		.02-.03		stems	
" "	n.a.		.08		leaves	sown April 1980
" "	Extremes of 90		.032-.082		whole plant	harvested Nov. 1980, vegetative
" "	Regression on 90		.044		irrigated	
			.062		non-irrigated	

3.2 Content of Essential Oils and Resin in Eucalyptus spp.

Analyses for E. globulus from California and other species from Australia are shown in Table 3-2. Nishimura and Calvin (1979) and Nishimura et al. (1980) reported analysis as per cent of fresh weight, and also gave a content of dry residue, and of water plus water solubles. As a first approximation it was assumed that total dry matter content was 1.5 times the residue content. This gave dry matter contents of 0.397 to 0.506 for the juvenile foliage, and 0.44 to 0.69 for the various parts of E. globulus, which appear reasonable. The proportion of essential oils ranges from 0.02 to 0.10 of dry weight, or 0.012 to 0.042 of fresh weight, which is consistent with commercial oil recoveries of 0.01 to 0.035 of fresh weight reported by Small (1977). The non-volatile part of the acetone extract is less variable, ranging from 0.09 to 0.11 of dry weight. Small (1977) reported oil contents for 15 commercial Eucalyptus spp. and E. radiata and E. dives were the only species to yield 0.03 to 0.035 of fresh weight as oil.

E. radiata and E. dives appear particularly attractive because of their high content of essential oils that can be used directly as liquid fuels. Further research is needed to identify the nature of the non-volatiles and to assess whether they can be readily converted to liquid fuels.

Table 3-2 Proportion of essential oils and total acetone extract in dry matter of Eucalyptus spp.

Source of data	Species and part	Proportion of essential oils	Proportion of total acetone extract
Nielson et al. (1977)	E. globulus dried (herbarium)		.12
Nishimura and Calvin (1979)*	E. globulus fresh fruit	.014	.060
	" fresh leaves and branches	.022	.115
	" fresh shoots	.028	.136
Nishimura et al. (1980)*	E. radiata fresh juvenile shoots	.100	.208
	E. dives " " "	.068	.178
	E. polybractea " "	.045	.134
	E. pulverulenta " "	.050	.164

* Assumed that dry matter content = 1.5 x residue content

3.3 Surveys of Resin-Containing Plants in the USA

A summary of the surveys of resin-containing plants is shown in Table 3-3.

Johnson and Hinman (1980) surveyed nearly 200 species from the drier parts of southwestern USA and concluded that E. lathyris was the only species that showed promise in terms of biomass production and content of resins.

Table 3-3 Summary of surveys of species in USA analysed for resins plus related compounds

Author	No. of species analysed	No. of species reported	Proportion of dry weight - acetone extract	Most promising species
Polhamus (1967) (T.A. Edison)	2000			Not yet sighted by us, seen by Calvin and Buchanan
Nielson _et_ _al_. (1977)		12.	.042 to .165 - whole plant	_Monadenium_ _rhizophorum_
			.25 - leaves	_Euphorbia_ _lathyris_
			.40 - seeds	" "
			Proportion of dry weight Resin I Polyphenols	
Buchanan _et_ _al_. 1978a	106 mostly Illinois	16	.022-.076 .04-.188	_Ambrosia_ _trifida_
Buchanan _et_ _al_. 1978b	100 Illinois	20	.020-.097 .035-.214	_Euphorbia_ _dentata_
Buchanan _et_ _al_. 1978c	Summary of 1978a,b	12 Illinois 2 California	.015-.097 .026-.247	_Euphorbia_ _dentata_ _Euphorbia_ _lathyris_
Johnson & Hinman 1980	nearly 200			_Euphorbia_ _lathyris_ only promising species

Buchanan *et al*. (1978a,b,c) examined 200 species from many families from Illinois and adjoining states and 8 species from other parts of USA, including *E. lathyris* and *Parthenium argentatum* from California. For most Illinois species samples were taken at least twice in the year, and some were sampled in two or three years. The content of resin I (plant oils) in relation to date of sampling are shown in Fig. 3-1. It was assumed that samples taken in one year were from one location (i.e. one ecotype) and lines have been drawn between data points in one year. There is a very clear trend of increasing resin content with sampling later in the season. In only one case, *Asclepias syriaca* sampled on September 9 and October 23, was there a reversed trend, and the authors have specifically mentioned the presence of mature seed in the sample for September 9. Apparently those seeds were shed before October 23. They also referred to presence of seed in three other samples, and those data points are marked with a dashed circle in Fig. 3-1. These all had resin contents appreciably higher than in other samples from those species. In some cases, e.g. Sassafras, there were marked differences in resin content in different years. If samples were taken in difference locations from year to year, that data might represent different ecotypes. Twelve species contain 0.12 to 0.25 of polyphenols, and the possibility of this fraction containing useful energy-rich components was discussed in Section 2.5. The sample of one year only guayule from California contained 0.04 of resin plus 0.07 of polyphenols giving a total of 0.11 extractable in acetone, which is in line with the acetone recovery from stems plus branches and roots of approximately 0.10 reported from pilot plant results with guayule harvested from wild stands in Mexico (Campos-Lopez *et al*. 1978). However the hydrocarbon content of 0.046 is very low compared with 0.12 recovered from wild guayule in the pilot plant in Mexico (Campos-Lopez *et al*. 1978) or the 0.20 that is considered feasible with improved varieties in US (Nivert *et al*. 1978).

The results of Nielson *et al*. (1977) appear high at first sight but it should be kept in mind that the acetone recoverable fraction includes resin I plus resin II or 'polyphenol' fraction, and that the samples are from herbarium specimens that would include flowers and/or fruits. The herbarium specimens may have proportions of leaf, stem, flower and fruit that may be different from the proportion obtained by mechanically harvesting a crop of that species, so these analyses must be used with caution. Five species had acetone extractables exceeding 0.1. *Cryptostegia grandiflora* had 0.133 which is very high compared with Australian data for cyclohexane extractables in Table 3-4. Other species exceeding 0.1 of acetone extractables were *Eucalyptus globulus*, *Monadenium rhizophorum*, *Sarcostemma viminale* and *Synadenium grantii*. All these species seem worthy of further examination, and initially that should be an examination of their polyphenol fraction to assess its suitability as a high energy component.

We have not seen the report of Edison's survey (Polhamus 1967), but it was cited by both Buchanan *et al*. (1978c) and Calvin (1978), and we might assume that they would have reported any species with high acetone extractables.

FIGURE 3-1

Proportion of resin 1 (plant oils) in relation to date of harvest for data of Buchanan et al. (1978c). Lines join data for samples for same year. Data points for samples with mature seed surrounded by a dashed circle.

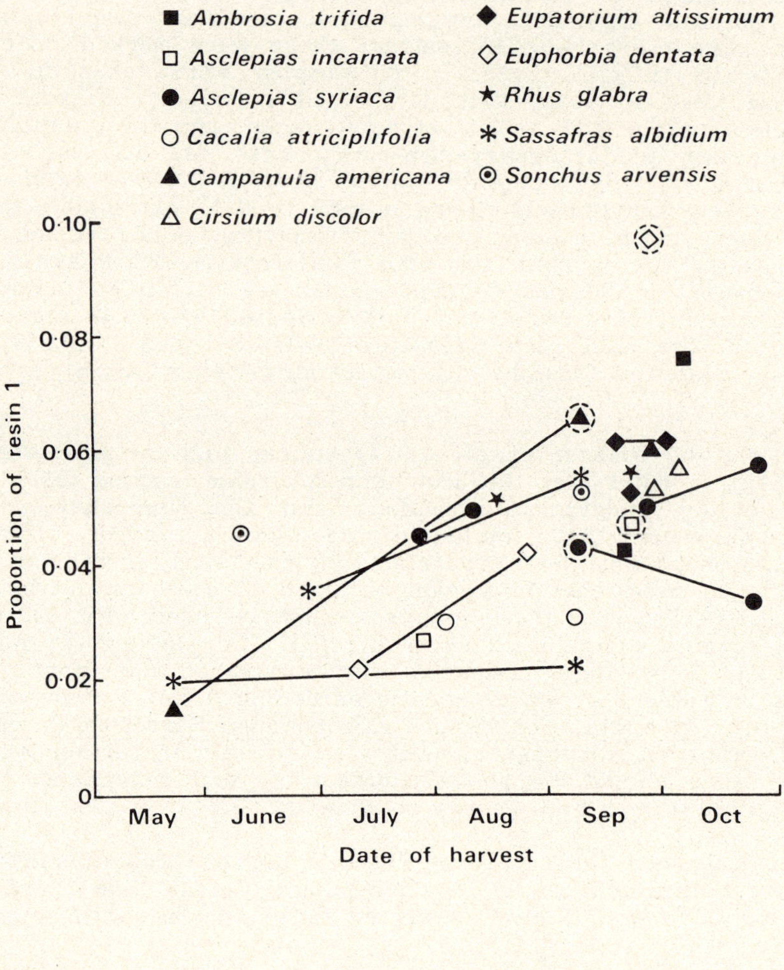

3.4 Investigations of Resin-containing Plants in Australia

Although resin-containing plants have received much publicity in Australian newspapers, only a very small amount of research has been carried out. The only quantitative data available is the content of resin I plus true hydrocarbons as shown in Table 3-4.

Table 3-4 Content of resin I plus true hydrocarbons in the dry matter of some latex-bearing plant species in Australia.

Plant species	Resin I + true hydrocarbons as % DW
Asclepias rotundifolia	10.0*
Araujia sericofera	4.7*
Plumeria rubra	6.1*
Euphorbia paralias	7.5*
Euphorbia tirucalli	4.3*
Calotropis procera	5.3* 5.8**
Calotropis gigantea	5.7**
Cryptostegia madagascariensis	5.2*
Cryptostegia grandiflora	5.7* 3.5**
Asclepias fruticosa	5.2**

* Ground to 0.1mm mesh and extracted with n-heptane (Tideman and Hawker 1981a)

** Ground to 1cm mesh and extracted with cyclohexane (Williams et al. 1981)

Growth of the following plants under field conditions is being tested at Roseworthy Agricultural College by Mr. F.J. van der Sommen.

Asclepias rotundifolia
Asclepias fruticosa
Euphorbia lathyris
Euphorbia tirucalli
Euphorbia barnardii
Euphorbia canariensis
Euphorbia pulcherrima (poinsettia)
Plumiera rubra (frangipani)
Parthenium argentatum (guayule)
Sarcostemma australe

Estimates of the hydrocarbon content of these plants are being obtained using the heptane extraction technique. Members of the Department of Chemical Technology of the S.A. Institute of Technology have carried out some screening trials with a few species of latex bearing plants available in large quantities in and around Adelaide. Larger scale extraction of hydrocarbons was generally carried out. The Shell Company of Australia Ltd. working at the CSIRO Division of Horticultural Research has briefly surveyed several species of plants for hydrocarbon content (Tideman and Hawker 1981a). Additional species studied to those mentioned were Euphorbia paralias (introduced weed), Calotropis procera, Cryptostegia madagascariensis and Cryptostegia grandiflora (all tropical introduced weeds from northern Australia). Williams et al. (1981) at Macquarie University have determined the hydrocarbon content of Cryptostegia grandiflora, Calotropis procera, Calotropis gigantea and Asclepias fruticosa. Most of these initial tests have been carried out on plants collected when and where available. It has not been possible to standardise conditions of growth of the plants. With this proviso in mind, it seems that Asclepias rotundifolia consistently shows the highest yields of heptane extractable material in the order of 8 to 10% of the dry weight. In vitro propagation at this species has been achieved (Tideman and Hawker 1981b). Cryptostegia grandiflora, the Queensland tropical weed, was investigated during the 2nd World War as a source of rubber (Dickson and Hartley 1946). Reports of high yields of hydrocarbons from Calotropis procera and Cryptostegia grandiflora grown in northern Australian have not so far been backed up by scientific data. Values for resin I plus true hydrocarbon content on a dry weight basis are less than 6% (Tideman and Hawker 1981a; Williams et al. 1981) and there is no data for biomass yield in the field. However 'Energy Agriculture' is funding research into biomass yield and hydrocarbon content of Cryptostegia grandiflora, Calotropis gigantea and Calotropis procera at James Cook University in Townsville. Natural stands of Cryptostegia grandiflora are being studied at Macrossan and Charters Towers, both inland from Townsville. Professor D. Griffiths is supervising the work.

Another company, Australian Biofuels Pty. Ltd., has seeded 20ha of flat land in the Ord Irrigation area near Kununurra with Calotropis procera. The seed was collected from local sources. Some plants were unirrigated, unfertilised and uncultivated while others had one irrigation and one cultivation. After six months growth, plants were about one metre high. Mr. Rex Jettner of Terri Co. Pty. Ltd. has sown plots of Calotropis procera 100m x 12m at Adelaide River outside of Darwin. Seed was collected near Katherine.

From the above trials it is obvious that Calotropis procera can be grown as a crop from seed by conventional agricultural techniques. Unfortunately no data on growth rates or plant spacing have been obtained.

The general conclusions of observations in the northern part of Australia are that Cryptostegia grandiflora and Calotropis procera, both well established weeds, regrow after being cut back, and that Calotropis procera, at least, can be grown from seed as a crop. Biomass yields and resin content under different seasonal conditions are not known.

Guayule (Parthenium argentatum) has been grown in Australia for various trial periods for many years. At the request of the Council for Scientific and Industrial Research, investigations commenced in July 1942 at the Waite Agricultural Research Institute in South Australia to assess the feasibility of producing rubber from guayule in South Australia. The local investigations were not continued after cessation of war. Because of the limited amount of time, labour and funds during the war, very few conclusive results were obtained although considerable progress was made on germination techniques and identification of the best soil and climatic conditions (Prescott 1943; Crocker and Trumble 1945; and Shapter 1952).

Various small scale plantings have taken place in Western Australian in 1929, 1943 and 1959. More recently, in 1977, CSIRO started small trials in Western New South Wales to determine the appropriate soil and climatic conditions for guayule. Dr. R. Downes of the CSIRO Division of Plant Industry is testing nine cultivars at three sites (Canberra, Narrabri and Charleville) but to date has only survival and yield data but no information on rubber content. Both the N.S.W. and S.A. Governments are becoming involved in trials with guayule in their States. N.S.W. is cooperating with the USDA Special Program on guayule and other plants and S.A. has approached the USDA recently with a view to cooperative research.

3.5 Conclusions

Our economic assessment (see Section 8.3) of the cost of production of resins indicate that crop species would need to contain 0.15 to 0.20 of total resins to be economically competitive with oils from oilseeds. To date no species have given contents of resin I greater then 0.10. Two *Eucalyptus* spp. have essential oil contents of 0.07 to 0.10 and non-volatile resins of about 0.10. *Euphorbia lathyris* was reported to have resin II contents in the range of 0.03 to 0.07. Several other species have high contents of 'polyphenol' fraction but in some species the energy content of that fraction is much lower than in resin II of *E. lathyris*. Further chemical studies are needed to identify the components of the 'polyphenol' fraction and to assess their potential for conversion to liquid fuels.

The review of past work indicates that crops closer to physiological maturity can have higher resin contents, and this aspect needs further examination for virtually all species. If growth was carried to full physiological maturity, the crops would be virtually oilseed crops.

4 CANDIDATE CROP SPECIES

Two species shown by US studies to be high in resin and/or rubber, and on which some research in Australia has been done, are considered as candidate species - Euphorbia lathyris (gopher plant) and Parthenium argentatum (guayule). In temperate Australia Asclepias rotundifolia has been shown to have a resin content of about 0.10 (Tideman and Hawker 1981a), and is also considered a candidate species. In tropical Australia the only adapted species are Calotropis procera, Calotropis gigantea and Cryptostegia grandiflora, but the resin content of all of these species is only about 0.05 (see Table 3-4). In the absence of better species, Calotropis procera is considered as the candidate species for tropical Australia. However, the estimation of yield and cost of feedstock are based on generalised information not specific to species. Thus production systems and costs for Asclepias fruticosa would not be different to that proposed for Asclepias rotundifolia, and the same would apply for Calotropis procera and Cryptostegia grandiflora.

Several Eucalyptus spp. have contents of essential oils plus resins in the range of 0.12 to 0.20 of dry weight, i.e. appreciably higher than the resin contents of the above candidate species. Further data are being sought, with the objective of making an economic assessment of the most promising species.

4.1 Euphorbia lathyris

E. lathyris is a biennial growing from 0.5 to 2 metres high. It is considered to be native to western mountains in China or the central and eastern Mediterranean region (Peoples et al. 1981). It has a long history of utilisation by man. The seeds were used as substitutes for coffee and for capers, as an emetic, and in the treatment of dropsy. The plant has occasionally been cultivated as an oilseed crop in parts of the Soviet Union, China and Japan. E. lathyris is widely distributed around the world in temperate mesic habitats.

In California two ecotypes have been identified. The northern ecotype, as used by Sachs et al. (1981) has a vernalisation or crop chilling requirement of six to eight weeks before it becomes reproductive in the spring and early summer. A similar feature was shown by E. lathyris grown at Melbourne University (Coffey and Halloran 1981). A southern Californian ecotype being grown experimentally in Arizona grows to maturity in about 180 days regardless of time of sowing, and it is more productive than the northern ecotype in Arizona (Peoples: personal communication).

E. lathyris is very prone to a range of diseases under high moisture stress when temperatures are high (Sachs et al. 1981; Peoples et al. 1981). At Davis, California, experiments with E. lathyris ceased after one season because the yield of high energy compounds was smaller than that achievable with current oilseed crops over a shorter time period, and because the scope for improvement by breeding appeared very limited (Sachs et al. 1981). In Arizona research is continuing, including an active breeding program (Peoples et al. 1981), because:

(a) the southern ecotype and related types could be grown to avoid the disease problem associated with high temperatures;

(b) there appears to be scope for improving resin content by breeding and selection. Content of cyclohexane extractables ranges from 3 to 16% and an additional lower energy content material, (excluding carbohydrates), probably similar to the ether solubles of Nemethy et al. (1980), is also produced in a range of 3 to 10%. They suggest a realistic yield of 15% of hydrocarbon-like materials; and
(c) they conclude that experimental crop densities have not been adequate to achieve optimum dry matter yields.

Field experiments at Davis, California (Sachs et al. 1981) and Tucson, Arizona (Peoples: personal communication) show no response to application of nitrogen fertiliser in either dry matter yield or resin content. Peoples et al. (1981) report that numerous plant samples suggest an average nitrogen level of 0.7%, which is very low by normal crop standards. The N/P ratio is low at about 3.5 compared with many crops of about 8. Both Peoples et al. (1981) and S. Coffey (personal communication) report marked responses to various levels of phosphorus in water culture experiments.

4.2 Parthenium argentatum (Guayule)

Guayule, a member of the Family Compositae, is native to northern Mexico and the northwestern corner of Texas. New cultivars developed in the USA are reported to contain approximately 0.2 of the dry weight as rubber and approximately 0.1 of their dry weight as resins (Nivert et al. 1978). This resin content is as high as that reported for other candidate crop species, and we therefore consider it an appropriate candidate crop. However, the economics of production of that resin will be greatly influenced by the high content of rubber which is currently about five times as valuable as crude oil.

In the 18th century Indians in Northern Mexico were seen playing with rubber balls for which the rubber was said to have been provided by chewing of the bark of guayule (McGinnies 1978). The first commercial extraction of rubber was carried out in New York in 1888. The first factory was built in Mexico in 1902, and rubber was produced commercially and mainly exported to USA (Fig. 4-1) up to 1951 (Velazquez et al. 1978). A total of about 120 000t of guayule rubber was exported to USA over this period. All of the guayule that was needed for that production was obtained by harvesting of natural stands of guayule. The native stands apparently regenerated successfully after harvesting in the three main periods of production (see Fig. 4-1). Since harvesting ceased in 1950 the natural stands have regenerated to the extent that existing resources of guayule were assessed at 2.6Mt in 1975 (Gomez-Contreras 1978).

Studies of the cultivation of guayule were commenced in 1910 by the Continental-Mexican Company. They established plantations in Arizona in 1916 and at Salinas, California in 1925 (Hammond and Polhamus 1965). Only about 1400t of rubber were produced in USA in the late twenties and virtually none in the thirties. In 1942 the US government took over all guayule in the Emergency Rubber Project. A massive program of research and development was undertaken, but only produced about 1400t before guayule production was terminated and existing guayule crops were destroyed in 1946, due to the availability of low-priced rubber from the rapidly growing synthetic rubber industry and from the Hevea plantations in southeast Asia (National Academy of Sciences 1977). However, research was continued by USDA until 1959 (Hammond and Polhamus 1965).

FIGURE 4.1

Exports of guayule rubber from Mexico (Velazquez et al. 1978)

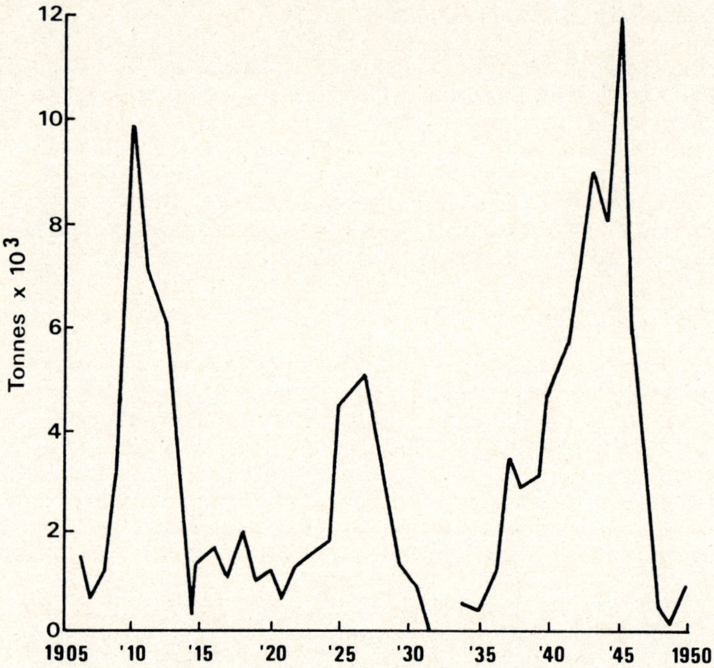

In 1973 the Mexican Government initiated a research and development program on guayule, mainly based at the Centro de Investigacion en Quimica Aplicada at Saltillo. A pilot plant to process one tonne of shrub per day was completed in 1976 at Saltillo. It employed new technologies, including Bauer mills for grinding the crop material and solvent extraction for recovery of resins and rubber, but it continued to use flotation to separate rubber from the bagasse (Campos-Lopez et al. 1978).

In 1975 the US National Academy of Sciences organised an Ad Hoc Panel to examine the state of knowledge and the future promise of guayule as a source of rubber (National Academy of Sciences 1977). The papers presented at the meeting of that panel in Arizona in November 1975 were distributed as the Proceedings of the First International Conference on Guayule. The second International Conference on Guayule was held at Saltillo, Mexico in early August 1977 (Consejo Nacional de Ciencia y Tecnologia, Mexico 1978), and further international meetings were held in 1979 and 1981.

Research on guayule recommenced in USA in 1976, and in 1978 the Natural Latex Commercialisation Act was passed by Congress. This Act authorised a $30 million program for research, development and commercialisation of guayule, and resulted in the formation of the Joint Commission on Guayule Research and Commercialisation. Major areas of research and development include:

(a) seed multiplication and testing in four states of the best cultivars developed by the Emergency Rubber Project;
(b) germ plasm collection in Mexico and Texas and plant breeding to incorporate desirable properties from that germ plasm bank into existing cultivars;
(c) determination of desirable cultivation practices including seed pelleting, use of herbicides for weed control, and use of bioregulators to increase rubber content; and
(d) improved methods of rubber recovery using solvent extraction methods.

The Office of Arid Land Studies at the University of Arizona, in collaboration with the Midwest Research Institute, has recently completed a comprehensive technological assessment of the commercialisation of rubber from guayule (Foster _et al_. 1980).

4.3 _Calotropis procera_ (Rubber Bush)

C. procera (Family _Asclepiadaceae_) is a perennial shrub growing to 3 metres high. It is native of India, Iran and Africa, and it is naturalised as a weed in northern Australia, occurring in Queensland from Cairns to Normanton, along the Roper, Katherine and Victoria Rivers in the Northern Territory, and along the Ord River in the north of Western Australia. Spread has apparently been by seed, which is produced in the late dry season (October). It is primarily a weed where native pastures have been overgrazed. It is only grazed by cattle when alternative feed is in short supply. It is not a problem on land that is cultivated periodically, but has spread on some Ord River farms that have not been cultivated for some time.

In India the plant is used as a source of several drugs, for strong fibre from the bark, and a source of kapok-like material from the floss attached to the seeds. The dried coagulate from the white latex that exudes from cut plant surfaces contains 53 to 85% resins and 11 to 23% of caoutchouc = rubber. The latex is a strong irritant to the skin and mucous membranes and contains three heart poisons (Wealth of India 1950). _C. procera_ is not grown as a crop in India, but is sometimes planted around houses or other buildings.

Calotropis gigantea does not produce seed in Australia but grows readily from cuttings. It grows around Darwin but is not a weed of significance.

4.4 _Asclepias rotundifolia_

The common name is broadleaved cotton bush, and it belongs to the family Asclepiadaceae, the species of which are herbs or shrubs usually with a milky juice. Two other important species of this family in Australia are _Sarcostemma australe_ (milk bush, tableland caustic bush), a leafless shrub native to Australia with thick succulent branches which grows in the drier parts of all states except Victoria and Tasmania, and _A. fruticosa_ (narrowleaved cotton bush), a weed introduced from South Africa. _A. rotundifolia_, another native of South Africa, is not common in the eastern states but grows vigorously under South Australian conditions and is extremely common as a weed in the hills and foothills near Adelaide. A. rotundifolia can be distinguished from A. fruticosa by its broad and oblong leaves and larger, thicker internodes making up the shorter, stouter stems.

A. rotundifolia is a perennial shrub about one metre high with stout, hairy spreading stems. The purplish-white flowers are arranged in umbels of 10-20. The ripe seeds have long and silky hair tufts hence the name, cotton bush. The plant can grow and survive on poor soils with 250mm rainfall but also grows on better soils with 500mm rainfall. It does not cause serious problems in crops or pastures. Little is known of its nutrient requirements but its growth in sclerophyll forests suggests it can survive without large amounts of available nutrients. The plants seem relatively free of disease as weeds, however the younger stems and leaves are often covered with aphids which check but do not kill the plants.

5 ENVIRONMENTAL LIMITATIONS TO CROP GROWTH IN AUSTRALIA

5.1 Introduction

Any evaluation of a potential crop for any purpose should be based upon realistic productivity estimates, i.e. expected yields of the desired product in terms of t(DW)/ha.yr. For the resin producing plant species this will have two components:

(a) total above-ground biomass in t(DW)/ha.yr; and
(b) per cent content of resin.

The assumption is often made that these two components are independent and that significant gains made in one or other component can be multiplied to give an expected product yield. Because the synthesis and storage of higher molecular weight resins has an additional cost for the plant, it is unlikely that significant gains can be made in one component without having some effect on the other component. This is not to say that product yields can not be increased very significantly through selection and breeding programs, but it is a warning against simplistic assumptions that involve multiplying together the highest available figures, obtained quite independently, for the separate components.

Before proceeding to an analysis of environmental limitations and estimates of potential productivity it must be stressed that, for all of the potential resin crop species discussed in this report, the data base is totally inadequate. Even where some dry matter measurements and resin analyses have been taken, some of these are from single plants or parts of plants grown under glasshouse or widely-spaced conditions in small field plots. The recent study of *Euphorbia lathyris* at Davis, California (Sachs *et al*. 1981) is an exception. Another exception is guayule (*Parthenium argentatum*), where growth, development and hydrocarbon yield data as well as large plot yields are available from the world war II program in the USA and in Australia. However, the absence of matching weather, crop, soil, management data associated with such yields, the lack of information on irrigation application, and the probably lower productivity of the winter rainfall environments in which these experiments were conducted makes interpretation of the research difficult. Any extrapolation of these data to other environments and/or totally different production systems must be regarded with due caution.

The need for productivity data gathered under carefully controlled and monitored conditions needs to be emphasised. When adequate production data have been generated through appropriate field experiments, further downward adjustments are necessary to extrapolate to farm yields (Davidson 1962).

The supposed arid adaptations of some of the candidate species have led to proposals for extensive plantations in areas currently marginal for annual cereal grains and/or beyond into the arid pastoral zones. This may be possible, but higher risks of poor crop establishment, low productivity, higher infrastructure costs and higher risks of total crop loss make this unlikely. If such crops are to find a place it is much more likely to be in competition with established crops in existing or potential arable lands. Given that the candidate species are assumed to be grown on an extensive, highly mechanized basis the same basic constraints of climate, terrain and soil as evaluated by Nix (1978) in estimating potentially arable areas in Australia should apply. If these potential new crop species are not to compete with existing food and fibre production, then significant areas of undeveloped land could be available in south-western Western Australia (3 Mha), south-eastern Australia (0.75 Mha), eastern Australia (17 Mha), Cape York (2 Mha), Northern Territory (1 Mha) and the Kimberleys of W.A. (.025 Mha). Obviously these regions span a wide range of climatic environments and different species will be required to exploit them.

5.2 Some Estimates of Potential Biomass Production in Australia.

In many of the proposals for development of resin-containing crop plantations, totally unrealistic assumptions of potential biomass and resin yields have been made. To provide some perspective on this point it may be useful to consider basic environmental constraints of light energy, temperature, water and nutrients. In the Australian environment light energy is rarely limiting, although the radiation/temperature balance may significantly influence carbon assimilation and storage and influence choice of crop species and/or cultivar. Temperature is often limiting in that temperatures above and below those optimal for dry matter production will be experienced at some time during the year. However, careful choice of appropriate crop/cultivar will minimise this constraint. Diurnal and season variation in temperature is most pronounced in southern and inland regions. Water is the dominant limiting factor in the Australian environment and the amount of water available for transpiration by the crop sets an upper limit to potential biomass yields. Nutrients, particularly phosphorus and nitrogen, limit yields in most soils in Australia, but towards the arid limits of dryland cropping, responses to added nutrients may be marginal and uneconomic. Where water is non-limiting, such as in humid environments and/or with irrigation, nutrient addition becomes a very important aspect of management.

A useful first approximation to biomass production at any location can be obtained from general empirical relationships between measurements of net primary productivity of both native plant and cultivated communities and various simple climatic indices. Thus, Lieth (1973) derived two such equations, one based on mean annual precipitation:

$$Y = 3000(1 - e^{0.000664x}) \qquad (1)$$

where Y = net primary productivity (g(DW)/m^2.yr)

x = precipitation (mm/yr)

and another based on mean annual temperature:

$$Y = 3000/(1 + e^{1.315 - 0.119x}) \qquad (2)$$

where Y = net primary productivity (g(DW)/m^2.yr)

x = mean annual temperature (°C)

In both cases, an upper limit of 3000g(DW)/m^2.yr (30t(DW)/ha.yr) has been set. Given that the data used were mainly from natural communities (but presumably very well adapted to the particular site) this limit and the derived equations might seem inappropriate. However, comparison with biomass yield from extensive, low to medium-input agriculture suggests that they are broadly comparable. Major divergence occurs only where high-technology, high-input systems are matched with near optimum thermal and water regimes.

For present purposes, application of Equation (1) to Australian precipitation data produces the patterns shown in Fig. 5-1. For the southern dryland cropping zones, predicted biomass yields range from around 5t(DW)/ha.yr at the arid margins to around 10t(DW)/ha.yr at the humid margins at the wheat zones. While wheat crops at the arid margin might only average 3 to 4t(DW)/ha.yr of above ground biomass, these are growing for part of the year only and do not equate with perennial or biennial crops. For the eastern zone the range is between 7 to 12t(DW)/ha.yr and in the northern zone between 10 to 20t(DW)/ha.yr. While temperatures generally are more favourable during the period when water is available in the east and north, higher evaporation rates tend to reduce the effectiveness of precipitation. Accordingly, a more appropriate function should be one that relates net primary productivity to cumulative evapotranspiration. Such an relationship has been developed by Rosenzweig (1968):

$$\log^{10} Y = 1.66 \log^{10} AE - 1.66$$

$$Y = 0.0219 \times AE^{1 \cdot 66} \qquad (3)$$

where Y = net primary productivity (g(DW)/m^2.yr)

AE = cumulative evapotranspiration (mm)

In this case, cumulative evapotranspiration has been calculated for some 1000 locations in Australia using a water balance that takes into account precipitation, evaporation and soil water storage. The resultant patterns shown in Fig. 5-2 are in broad agreement with Fig. 5-1.

For a crude estimate of dry matter yields with water non-limiting (i.e. irrigated cropping) we can employ Equation (2). Again, using data for some 1000 stations in Australia, the resultant pattern is shown in Fig. 5-3. This suggests that irrigated dry matter yields in the southern zone could be 18 to 20t/ha.yr, in the eastern zone 20 to 25t/ha.yr, while across the northern zone yields should exceed 25t/ha.yr. However, remember that the function is based on perennial vegetation, while many irrigated crop systems have significant periods of the twelve month cycle when land is idle or when young crops have not developed a full canopy for maximum photosynthesis. Taking such factors into account, these estimates do not seem at all unreasonable and can provide useful approximate estimates of likely dry matter yields. At the least, they can provide a framework against which the more extreme and inflated estimates of biomass yield can be compared.

FIGURE 5.1

Estimated net primary productivity (t(DW)/ha.yr) based only on precipitation (Leith 1973).

$$Y = 30\ (1-e^{-0.000664x})\ \text{where } X = \text{Precipitation (mm/yr)}$$

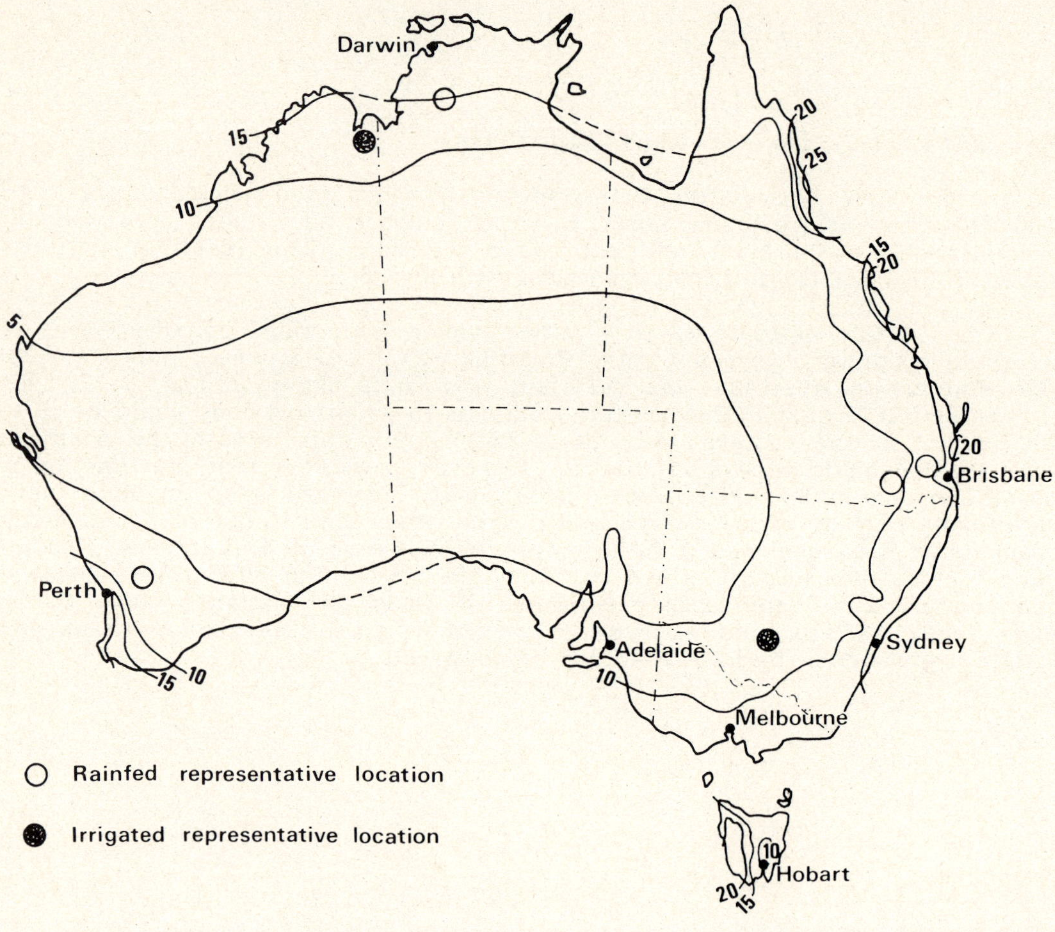

FIGURE 5.2

Estimated net primary productivity (t(DW)/ha.yr) based on cumulative evapotranspiration (Rosenzweig (1968).

$$Y = 0.0219 \times AE^{1 \cdot 66} \text{ where } AE = \Sigma \text{ evapotranspiration (mm)}$$

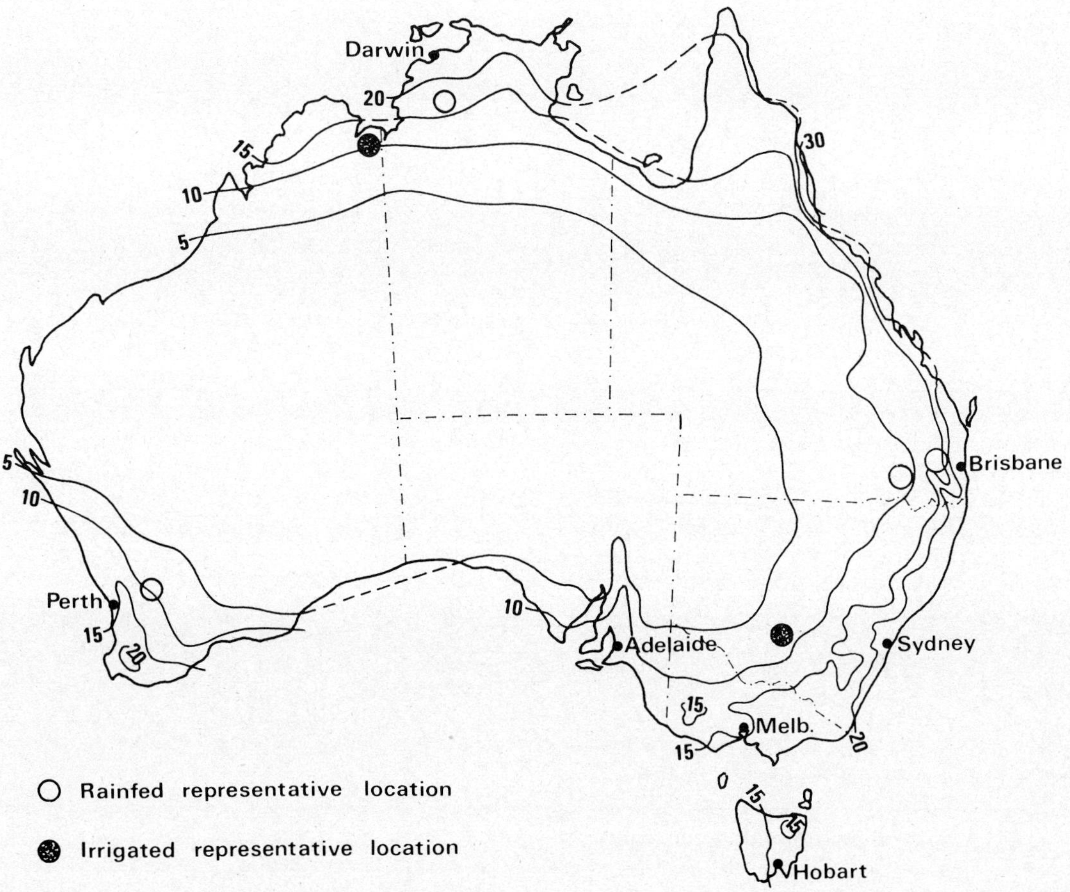

FIGURE 5.3

Estimated net primary productivity (t(DW)/ha.yr) based on mean annual temperature with water not limiting (Leith 1973). This is equivalent to fully irrigated cropping.

$$Y = 30/(1 + e^{-1.315 - 0.119x}) \text{ where } x = \text{mean annual temperature } (^{o}C)$$

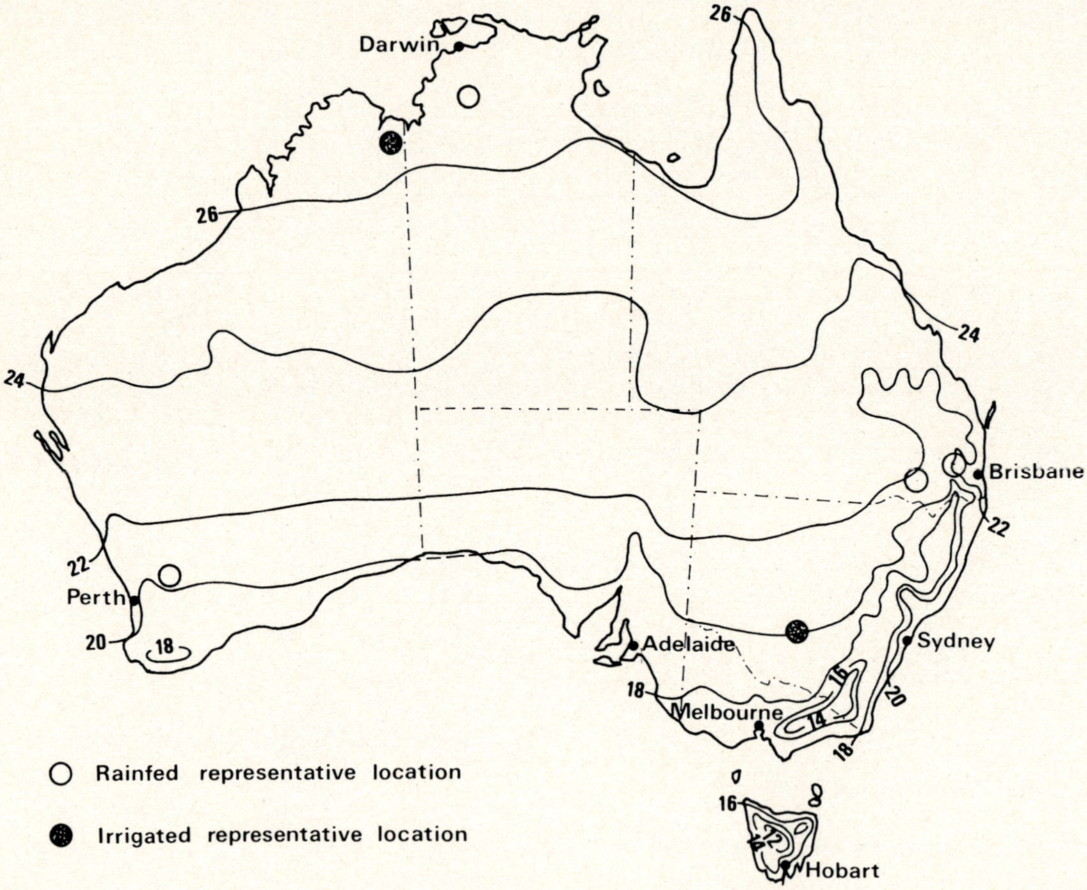

6 POTENTIAL FARMING SYSTEMS

Four of the candidate species identified earlier in this report, have been selected for more detailed analysis, as potential resin crop components of various cropping systems. Three species, Euphorbia lathyris, Asclepias rotundifolia and Calotropis procera are identified as appropriate dryland crop species for the southern, eastern and northern agroclimatic regions respectively. In addition, Euphorbia lathyris and Calotropis procera are considered as irrigated crops in the southern and northern regions respectively. The fourth species, guayule Parthenium argentatum is a special case since it is a source of high-value hydrocarbon (rubber) as well as significant amounts of resin.

6.1 Southern Agroclimatic Zone

Although most of the potentially arable land in this zone has been developed for crop and sown pasture, almost 5Mha remain undeveloped. Competition from alternative land uses and the relatively dispersed nature of parcels of arable land in the higher rainfall sector of this zone militate against development of significant large areas of resin crop production. The bulk of the remaining undeveloped land (3.75Mha) lies in the drier inland sectors of the wheat-sheep zone, with most of this (3Mha) in Western Australia. Accordingly, this sector is seen as the most likely prospect for resin crop development.

The winter wet, summer dry, rainfall pattern throughout this southern wheat-sheep belt is not conducive to perennial crops, although adapted, deep-rooted species may survive the hot, dry summers. Indeed, such a crop could play an important role in reducing the throughput of water to the water table in salt affected catchments. Euphorbia lathyris could be selected and bred for later maturity, thus maximising the amount of water withdrawn from the profile. Limited field evidence from Roseworthy, S.A. (van der Sommen: personal communication) suggest that this species is capable of continued growth and extraction of deeper soil water well into the hot dry summer period. Agronomic data under both irrigated and dryland conditions at Davis, California, has been reported by (Sachs et al. (1981).

Two possible systems are examined:

(i) dryland (representative location in eastern wheatbelt, W.A.); and
(ii) irrigated (representative location in M.I.A., N.S.W.).

In each case, all cultivation and seeding operations would be carried out with conventional wheat farming equipment. Harvesting and baling would be carried out with currently available haymaking equipment.

6.1.1 Dryland Annual Crop

It is assumed that the resin crop is incorporated into the existing farming system and that it would become part of a rotation and replace one third of farm area sown to crop in any one year. Sowing with the normal combine with alternative seed tubes blocked to give a row spacing of 35cm and at a rate to give 7.5cm between plants in the row, should favour initial establishment and development of this radially symmetrical plant and allow branching to close over the row later in the growing season. Obviously plant population and spacing will demand fine tuning for specific environments in order to maximise light interception and efficiency of water use over the entire crop cycle.

Early autumn sowing will be essential as lower soil temperatures are not conducive to germination and establishment. Data provided in Sachs et al. (1981) permit estimation of a threshold temperature for germination of about 7°C, compared with that for wheat of 2.6°C, or barley of 2.6°C, or oats of 2.2°C (Angus et al. 1980). It is more comparable with sunflower (7.9°C) or safflower (7.4°C). Optimum temperatures for germination exceed 20°C. Accordingly, sowing on the earliest possible rains in autumn is likely to be a critical part of cropping strategy. Rapid germination and early development is necessary to reduce weed competition. Although Sachs et al. (1981) reported that they had not identified a suitable herbicide, it is assumed that a post-emergent spray will be necessary to control grass weeds, even with early sowing.

The cool adapted selection of E. lathyris used by Sachs et al. (1981) was shown to have a chilling requirement. This could be a disadvantage in the colder parts of the southern wheat belt since receipt of the chilling requirement during the winter would lead to flowering and maturity early in spring. However, this could be a necessary requirement if E. Lathyris was to be grown as a seed crop. The southern Californian ecotype that does not have a chilling requirement would be satisfactory as a vegetative crop. Prolonged vegetative development of this tap rooted species should ensure continued water extraction from deeper soil layers and so it is assumed that leaves remain green and active well beyond the wheat harvest. However, harvesting must proceed before any substantial leaf fall or leaf senescence. Even with a deeper root system it seems unlikely that the crop could remain green much more than 1 to 2 months beyond maturity and harvest of the cereal grain crops, particularly on the soils being cropped.

Harvesting would proceed on the basis that maximum retention of leaf is necessary. Mowing and hay-conditioning are necessary for rapid drying of thicker stems in the field. Resultant windrows are picked up and baled in large round bales for storage in the field or elsewhere on the farm.

Yield estimation is difficult. Direct comparison with existing cereal crops would suggest that dry matter yields of between 3t/ha at the drier margin and 10t/ha at the wetter margin of the southern wheat belt should be attainable. These values tend to be lower than those derived by either the Lieth (1973) or Rosenzweig (1968) equations. However, those equations are based upon dry matter production of perennial vegetation. The longer growing season of E. lathyris, coupled with potential water extraction from deeper soil layers may well lead to higher biomass yields than that of associated cereal crops.

Sachs et al. (1981) provide data for water use and dry matter yield of an irrigated crop at Davis, California that indicate a water use efficiency of 25 to 30 kg(DW)/ha millimetre of available water and this is comparable with rainfed wheat crops in southern Australia. On a probability basis, the current wheat zone has between 150mm and 400mm available water during the crop season in eight years out of ten. On water use efficiency of 25kg/ha.mm, then dry matter yields at the limits range between 3.75 and

10t/ha in eight years in ten. The zone chosen for economic analysis has a crop available water of 200mm in eight years in ten indicating dry matter yields at 5t/ha. With adjustments for expected farm yield (i.e. 66% of expected yield) we get a rounded estimate of 3t/ha.

6.1.2 Irrigated Annual Crop

It is assumed that the resin crop can be grown either as a cool season replacement for irrigated wheat crops or as a warm season replacement for irrigated summer grains such as maize or grain sorghum. Data from Sachs et al. (1981) suggest that at Davis, an environment similar to that at Griffith, N.S.W., maximum yields are likely to be obtained from a summer grown crop seeded in spring. These authors report yields of from 16.3 to 19.3t(DW)/ha from an eight month crop optimally supplied with water by spray irrigation. These are small plot experimental yields from very fertile soils and with hand cultivation for weed control. On this basis, and by comparison with the Lieth (1973) temperature based equation, which indicates a 15t/ha dry matter yield for an eight month crop at Griffith we might expect that commercial irrigated crops of E. lathyris of some eights months duration would approach 10t/ha (66% of predicted yield).

Although similar cultivation treatments to that of the dryland crops are postulated, additional land forming operations may be necessary for irrigation. Fertilizer requirements are unknown but we can assume that phosphate and minor nutrients will be necessary on deficient soils.

6.2 Eastern Agroclimatic Zone

By far the largest area of undeveloped, potentially arable land is in this zone. Again, the more humid coastal sectors are eliminated because of competition from a wide range of land uses and the highly dispersed nature of potentially arable parcels of land. Also, high summer rainfall and relative humidity predisposes crops to fungus disease and militates against harvesting operations based on field drying. The bulk of the remaining undeveloped land (17Mha) lies inland from the coastal ranges. In the southern, cooler sector (7Mha undeveloped), extensive dryland and irrigated cropping development is in progress. Winter cereals predominate, but dryland summer crops such as grain sorghum and sunflower and irrigated cotton and soybean are important. In the northern, warmer sector (10Mha undeveloped) both summer and winter cropping are important, but dryland and irrigated farming systems are still in a pioneering stage of development. Throughout the eastern zone, the reduced seasonality of the water regime suggests that a perennial crop species could be most relevant. The candidate crop selected is the perennial milkweed, Asclepias rotundifolia.

In this case only a dryland cropping system is considered, since irrigation water is used for the production of high-value crops such as cotton, soybean and vegetables. Again it is assumed that the resin-containing crop is integrated with existing cropping patterns and that it replaces one third of the farm area sown to crop in any one year. Because the bulk of the crop's dry matter will be produced over the warmer summer months it is assumed that it will replace a summer crop such as grain sorghum.

Germination of seed under laboratory conditions is very rapid and successful at 25°C. The small, compact, smooth-coated seeds should present few problems with conventional seeding equipment, whether of combine or row-planter type. Plant population and spacing cannot be specified, but a row-spacing wide enough to permit at least one interrow cultivation for weed control may be necessary. Herbicides will be necessary for control of vigorous summer grass weeds, but no information is available on this important component of the proposed farming system.

Because of the relatively high temperature requirements for germination and active growth, sowing date is assumed to be in late September-October, perhaps following a summer grain or oilseed crop. This would permit a small carry over store of soil water from winter rainfall and tide the developing crop through the inevitable dry spells during the early summer period. Harvesting of the crop could proceed over the period from August to December (10 to 14mths growing period) using techniques similar to those already described for E. lathyris. However, field drying would become progressively more hazardous later in the year due to early summer storms.

While Asclepias rotundifolia appears to be relatively disease free as a weed species, it is a host for the caterpillar of the well known wanderer butterfly Danaus plexippus and the lesser wanderer Danaus chrysippus. Some form of insecticidal control may prove necessary, if this resin crop species is to be grown in large areas.

Like most weeds, A. rotundifolia commonly occurs in disturbed habitats such as road verges, heavily grazed pastures, bare fallows, erosion gullies and waterways. No data exists for biomass yield, but assuming a 10 to 14 month growth cycle in the western Downs region, centred on Tara, and from the Lieth and Rosenzweig estimates a yield of from 8.3 to 11.6t(DW)/ha, when adjusted for commercial yield (x .66) we might expect yields (rounded) of from 6 to 8t(DW)/ha.

6.3 Northern Agroclimatic Zone

With the exception of a relatively small sector of high rainfall coastal and tableland environment on the east coast of Queensland, dryland cropping is virtually non-existent. Absence of established cropping systems makes any evaluation of arable land potential that much more speculative. Given known technology, Nix (1978) estimated that only about 3 Mha could be considered as potentially arable. Of this, Cape York had at least 2 Mha, the Northern Territory slightly less than 1 Mha and the North Kimberleys only .025 Mha. However, if suitable crop species and/or appropriate management could be developed for very large areas of presently intractable duplex soils and seasonally waterlogged grey earths, then upwards of 20 Mha could be available on Cape York and perhaps 3 Mha in the Northern Territory. Much of this land might well be suitable for perennial shrubby hydrocarbon crops.

The prevailing climatic pattern throughout this zone is summer wet - winter dry. Annual grasses and herbs are prominent, but perennial trees and shrubs dominate most landscapes. Three potential hydrocarbon crop species could be considered and all are perennial. These are the rubber vine (Cryptostegia grandiflora) and the shrubs, Calotropis procera and Calotropis gigantea. Some preliminary observations and measurements of growth of Cryptostegia grandiflora are being made at James Cook University by Mr. Ian

Carruthers under the supervision of Professor Dillwyn Griffiths. Despite the apparent rampant growth of this species, dry matter production from existing weed stands may not be as high as expected. This may prove to be equally true of both Calotropis species. Until field plantings at high plant populations are carried out and yield of biomass monitored from successive cuts, very little can be said about potential hydrocarbon yields under commercial conditions.

For the purpose of this exercise the hydrocarbon crop chosen is Calotropis procera. This is a perennial shrub now widely distributed as a weed across northern Australia. Examination of trial crops planted in the Ord Irrigation Area by Mr. Terry Mytton for Australian Biofuels Pty. Ltd., suggested that C. procera should present few problems using conventional equipment and standard farming practices. Obviously weed control will be essential, whether this be through cultivation between wider rows or through use of selective herbicide. With a row spacing of 75cm and an estimated population of 35 000 plants/ha, dry matter production at the end of the 1980-81 wet season was unimpressive and grass weeds were rampant. However, another plot that had one weed control cultivation and one irrigation and an estimated population of 250 000 plants/ha was more impressive.

Given high populations, weed control and added phosphate and nitrogen fertiliser, experimental yields of 15t(DW)/ha.yr and commercial yields of 10t(DW)/ha.yr (experimental yield x 0.66) should be attainable at dryland locations such as the Douglas-Daly development site in the Northern Territory and at least twice that yield i.e. experimental yields of 30t(DW)/ha.yr and commercial yields of 20t(DW)/ha.yr with irrigation on the Ord Irrigation Area. This can be compared with peak yield of the C4 crop sugarcane grown under pilot commercial conditions on the O.I.A. where dry matter yields exceed 50t/ha.yr and actual sugar yield can exceed 20t/ha.yr.

Harvesting of the biomass from Calotropis plantations can proceed for some 6 to 9mths, throughout the dry season. Mowing and baling is one option, but forage-harvesting with field bins may be more suitable for intensive irrigated production. Much more data are needed on variations in resin content in different parts of the plant and throughout the seasonal cycle. Timing and effect of cutting on regrowth pattern is another vital consideration.

6.4 Potential for Guayule (Parthenium argentatum)

Guayule (Parthenium argentatum) is a perennial, short, branching shrub native to the arid and semi-arid zones of northern Mexico and a small area in the 'Big Bend' of Texas, USA. Unlike the rubber in Hevea brasiliensis and other latex plants, guayule is contained within single, thin-walled cells in the stems, branches and roots. Leaves contain no rubber.

Field experiments, both dryland and irrigated, were conducted during and just after world war II, both in the USA and in Australia. Despite the fact that the guayule is native in lower latitudes with summer rainfall dominant, practically all of these experiments were conducted in winter rainfall dominant, Mediterranean type environments. Results in terms of both dry matter and rubber yields were not impressive, but significant improvements were made in agronomic methods, cultivars, and in processing technology. A classical agroclimatic evaluation of the Australian homoclime of guayule suggested that optimal climatic environments were rather to the north of the then current field experiments (Prescott 1943). Coonamble, N.S.W. was shown to be a close homoclime of Fort Stockton, Texas at the cold northern edge of the natural distribution.

Building on this earlier study, Nix and McMahon (unpublished) used a computer-based crop evaluation system to explore the homoclimes of guayule in Australia and to predict optimum locations for potential rubber production and reported these results at the First Australian Jojoba Conference in Bathurst, N.S.W. in 1979. These results indicate that optimum climatic environments were further to the north than earlier work had suggested and that the Mediterranean type environments favoured by both wartime and current field experimentation were unlikely to be highly productive even with irrigation. The central problem is that the temperature response of guayule leads to near complete cessation of growth during the colder months of these environments. On the other hand, cooler periods with high radiation levels are needed to promote a favourable carbon balance and rubber accumulation.

Because no relevant biomass and rubber yield data were available it was necessary to construct a dimensionless index for growth response, coupled with a calculated photothermal quotient set at a threshold of 1.3MJ/m^2.day-degree. This provides an index of potential carbon fixation per unit of developmental time and was used in this case as a measure of rubber production. Resultant values were scaled on a 0 to 10 continuum to provide a comparative measure of guayule potential. The resultant map showed a very close agreement with the homoclime map. On the criteria used, much of the east coast has values exceeding 9, but competing land uses and the threat of disease in the high humidity rule out these areas. Inland, the optimum zone (> 8) stretches from north of Charters Towers south to Goondiwindi and then a narrow zone bordering the New England Tablelands and extending down the Hunter Valley to Singleton. Further inland a zone with values of 7 to 8 extends from Hughenden in north Queensland south through Longreach, Charleville, Bolton, Bourke to Dubbo. In Western Australia, values exceed 7 along the narrow coastal strip from Esperance to Albany and from Bunbury to north of Geraldton.

The results of the experiments in California and Arizona suggested that guayule was intolerant of wet, waterlogged conditions and that deep, well drained loams were ideal. This would appear to rule out the most extensively cropped soils in the eastern agroclimatic zone, the cracking clays. Certainly, the finely cracking, self mulching clays may not prove suitable, but the grey and brown clays of the brigalow zone may do so. Very extensive areas of deep, lateritic, red earths occur throughout the eastern agroclimatic zone and these should prove eminently suitable for guayule. High intensity summer rainfall with a row crop makes for a potential erosion hazard. Whatever the soil, some form of contour layout and a bed and furrow system is likely to be necessary to minimise soil erosion and to remove excess water after heavy rains.

Although previous experiments with guayule concentrated on production of nursery seedlings and transplanting, it is assumed here that direct seeding will be possible. With the higher soil temperatures of the eastern agroclimatic zone, germination should be rapid and all the problems associated with sluggish germination and early growth minimised. However, precision seeding equipment such as that used for grain sorghum, sunflower, soybean, in this region is likely to be essential. The addition of moisture seeking planter tines and water injection equipment may reduce risks of poor establishment.

Population and spacing are critical components of cropping strategy. Here it is assumed that guayule is grown as a row crop, permitting interrow cultivation. Control of vigorous C4 grass weed species in summer may require the use of herbicides, but this will require further experimentation. On the finer textured soils, phosphate fertiliser may not be necessary but 40kg N are applied pre-sowing and again after the first harvest. On the lateritic red earths, it is likely that application of phosphorus and, in some cases, potassium will be necessary.

No yield data exist, but accepting the Lieth and Rosenzweig estimates, both around 10t(DW)/ha.yr and adjusting downwards to commercial yield (0.66) and again because of the relatively high energetic costs of the plant hydrocarbons (0.7) we get an estimated 4.7t(DW)/ha.yr, which is rounded to 5t(DW)/ha.yr. The initial harvest involves cutting the tops at a level of some 5cm above ground (i.e. pollarding), and baling. Guayule has been shown to regrow from such cut stems, but clearly more research will be necessary to establish the conditions under which such regrowth can occur. Here it is assumed that regrowth will occur and, following an interrow cultivation and application of 40kg/ha N, the crop is grown for a further two years. It is expected that growth rates will be higher because of root reserves and at the final harvest the roots in the upper 20cm are cut and lifted since they contain substantial amounts of rubber and resin. Because of these factors the yields in the second two year cycle are assumed to be equal to those of the initial three year cycle.

The estimated yield of 5t(DW)/ha.yr is high compared with the estimated US yield of 2.7t(DW)/ha.yr for dryland crops and is close to their estimate of 5.6t(DW)/ha.yr for irrigated crops (Foster et al. 1980). In view of the lack of Australian yield data, yields of 2 and 3 t(DW)/ha.yr will also be used in estimating the cost of production of guayule.

While it is possible to envisage complete guayule plantations in drier marginal areas currently not being cropped, guayule may well have a place in rotation with other crops in existing cropping areas. Current short season crops do not extract water much below 2m and this permits a build up of shallow groundwater. This process could pose a long-term threat to many sectors of the eastern agroclimatic zone since often significant amounts of salt (NaCl) are found deep in the profile. A five year guayule crop could do much to redress the balance disturbed by clearing the original deeply rooting tree vegetation.

Pests do not seem to present any particular problems, but pathogens could do so. Phytophora root rots, dieback and wilts have been noted, particularly under waterlogged conditions. This appears to be the primary limitation to guayule production in more humid environments and on more finely-textured soils. Again, properly surveyed layouts, contour alignment and bed and furrow systems may overcome such problems.

7 COSTS OF CROP PRODUCTION

Based on the considerations of farming systems described earlier, this section presents preliminary cost estimates of crop production at six representative sites around Australia in 1980-81 dollars. The costs include all costs for providing material year round to a plant processing 300t of resin crop per day and the sites represent a variety of environments, physical and economic, in which the resin crops might be grown. The estimates are best estimates from wide, plausible ranges. Appendix 1 shows details of the assumptions made in the costs of production at one site to illustrate the level of uncertainty present.

One technique of costing assumes that the resin crop replaces an existing crop, either wheat or sorghum. The revenue from the new crop then must meet the contribution of the old crop to farm overheads (its 'gross margin'), and must cover its own 'direct' cost. This assumption means that any increase or decrease in labour or machinery hours brought about by the substitution are costed only at the variable costs of machinery - fuel, oil, repairs and maintenance. The resin crops mostly appear to require less labour and machinery than wheat or sorghum and, as a result of the assumed costing techniques, no credit is taken for less labour used. For the two sites in Northern Australia where cropping is not developed, full farm costs are presented. Both the 'gross margin' and 'full cost' method are presented for guayule production.

7.1 Annual Resin Crop (Euphorbia lathyris) in the Southern Winter Rainfall Zone

The costing is based on the analysis of Eckersley (1979) of a 'typical' farm of the eastern wheatbelt of Western Australia. The farm size is large at 2600ha with 1200ha of arable land sown mostly to wheat; the soil type is light and suited to 'minimum' tillage practice; the resin crop is sown in April/ May with large tractors and replaces one third of the existing crops (400ha).

Yield is estimated to be 3t(DW)/ha and the crop is harvested after eight months in December/January. This yield is low in comparison with other sites that have been considered. The crop is mowed, windrowed and dried in the field, baled in large round bales and stored in the field. It is transported to the processing centre throughout the year. The costs are presented in Table 7-1. The major items are harvesting and the gross margin foregone in wheat production.

Table 7-1 The costs of growing an annual resin crop (Euphorbia lathyris) in the southern winter rainfall zone as part of an existing cropping system. Yield is assumed to 3t(DW)/ha.year.

	$/ha
1. Seed	6
2. Fertilizer 9kgP/ha as super @ $100/t super	10
3. Sprays - post emergent spray	15
- no insecticide	
- no fungicide	
4. Tractor operations (fuel, repairs and maintenance, oils and greases)	
(i) Plough @ 0.13 hrs/ha (130 pto kW)	
(ii) Cultivate @ 0.10 hrs/ha (130 pto kW)	
(iii) Fertilizer, sow @ 0.13 hrs/ha (130 pto kW) April, May	
Sub total 0.36 hrs/ha @ $10.50/hr	4
(iv) Spray @ 0.14 hrs/ha (45 pto kW) @ $5/hr	1
5. Harvest (December, January) Mowed, windrowed, field dried, baled in big round bales, stored infield and available year round. $20/t(dry), 3t/ha	60
6. Transport costs - 17km average haul 20¢/t.km. 3t/ha	10
7. Gross margin foregone in wheat production (1.1t @ $100/t less variable cost $44/ha)	66
	178
Cost of resin crop delivered ($/t) (3t(DW)/ha.yr)	59

7.2 The Costs of an Annual Irrigated Crop Grown in Southern Australia (Euphorbia lathyris) at Griffith, N.S.W.

The system assumed that Euphorbia lathyris replaces wheat at Griffith, N.S.W. Cost data were obtained from Davies et al. (1980) for the Dubbo region but these are considered applicable to Griffith. The system is one where a high yield of wheat 3.7t/ha is replaced by a yield of Euphorbia lathyris of 10t(DW)/ha.yr. Costs are presented in Table 7-2. The higher yield of resin crop and high gross margin of wheat displaced compensate for each other to produce a final cost comparable to others in this series.

Table 7-2 The costs of an annual irrigated crop of Euphorbia lathyris in southern Australia

	$/ha
1. Seed	12
2. Fertilizer; 30kg/ha P, 160kg double super @ $167/t	26
3. Tractor costs; 4hrs @ $6.60/hr	26
4. Water; 4.6M litres @ $2.50/M litres	12
5. Herbicide	15
6. Harvesting; contract @ $20/t, 10t/ha	200
7. Transport 8km @ 20¢/t.km, 10t/ha	16
	307
Gross margin on wheat foregone	202
Total Costs	509
Cost of resin crop delivered ($/t)(10t(DW)/ha.yr)	51

7.3 The Costs of Growing a Perennial Resin Crop (*Asclepias rotundifolia*) in the Region of Eastern Australia which received both Summer and Winter Rainfall

The model region taken is around Dalby in Queensland and costs are based on an analysis made by Thew (1980). The resin crop is assumed to replace summer grown sorghum and 7.5t(DW)/ha of resin crop is assumed to replace 3t/ha of grain sorghum. Farms in the region are taken as 500ha average size with 300ha in crop and it is assumed that the resin crop will replace one third of the traditional crop.

In this example the costs of seed bed preparation, sowing, and seed are spread over six years by calculating an equivalent annual cost at a 13% interest rate.

The costs of Table 7-3 are dominated by the gross margin the farmer now receives for sorghum. The price received by the farmer for the resin crop must be approximately $60/t to cover the contribution to overheads he now receives from sorghum of $80/t(sorghum) plus the variable costs associated with the resin crop.

Table 7-3 Production cost in the the eastern, summer and winter rainfall zone for *Asclepias rotundifolia*

	$/ha
1. Seed	2 *
2. Fertilizer	-
3. Sprays	-
4. Seedbed preparation	4 **
5. Cultivation	3
6. Herbicide	15
Aerial application	4
7. Harvest (7.5t/ha, $20/t)	150
8. Transport costs 7.5t/ha, 9km @ 20¢/t.km	13
9. Gross margin foregone in sorghum production	240
	437
Cost of resin crop delivered ($/t)(7.5t/ha.yr)	58

* Equivalent annual cost of $12/ha, see text
** Equivalent annual cost of $20/ha, see text

7.4 The Cost of a Perennial Resin Crop (*Calotropis procera*) Grown in a Northern Summer Rainfall Zone

Large scale cropping in the Northern Territory is not yet established and costs must be calculated on 'full cost' basis rather than a 'variable cost plus gross margin of displaced crop' basis. Because of the absence of large-scale cropping in the region the costs presented here are more speculative than others.

The system is based on a budget prepared by Whereat (personal communication) for 500ha of sorghum at Douglas Daly. The hydrocarbon crop is assumed to give yields of 10t(DW)/ha which is the highest 'dryland' yield we consider. Some discounting of current prices occurs in Table 7-4, e.g.

the current price of super-phosphate in Darwin is \$370/t which might be compared with the price used in southern Australia of \$100/t. In Table 7-4 a price of superphosphate of \$200/t is used which is the current Brisbane price plus freight from Brisbane to Darwin. In estimating average crop haulage distance it has been assumed that Calotropis procera can be grown continuously on all the cropping area available.

Table 7-4 The costs of growing a perennial resin crop (Calotropis procera) in the northern summer rainfall region of Australia

Variable costs	\$/ha
1. Seed \$12/ha (equivalent annual cost)	2
2. Fertilizer	
250kg super @ \$200/t (fertilizer price = Brisbane price plus freight (\$100/t))	50
63kg urea @ \$360/t	23
3. Pre-emergent herbicide	15
4. Tractor operation	
1 chisel plough, 3 scarifying sowing, \$22/ha; equivalent annual cost	5
5. Aerial application of herbicide	20
6. Harvesting 10t/ha \$20/t	200
7. Transport costs 4km @ 20¢/t.km x 10t/ha	8
	323
8. Fixed costs	
Labour, required for sowing 100ha/yr 1 man year @ \$15 000 p.a.	30
Machinery depreciation (prime cost \$160 000)	50
Interest on machinery capital	18
Miscellaneous overheads including accountancy, travel	20
Interest on land 500ha @ \$150/ha @ 13% p.a.	20
	461
Cost of resin crop delivered (\$/t)(10t(DW)/ha.yr)	46

The major points highlighted in this analysis are the potential for competitive production in the north if high yields are achieved and the high level of uncertainty surrounding the prices of individual items. Additional costs will be incurred because the area is undeveloped and while these are primarily infrastructure development costs such as roads, the remoteness of the location will also add to the cost of setting up the processing plant.

7.5 The Costs of a Perennial Resin Crop (Calotropis procera) Grown in Irrigated Northern Australia

The estimates are based on the budgets of Robertson (1980) for the Ord River Irrigation Area. They represent the most intensive farming system considered and yields expected are 20t(DW)/ha.yr. The crop is assumed to be Calotropis procera. The Ord River Irrigation Area is largely undeveloped and the costing technique used is a full cost analysis of a 240ha, two man, irrigated farm. The irrigated Ord offers the potential of harvesting the crop over a nine month dry season but the value of this is unknown without further knowledge of how resin content varies with season and how the crop fares when stored in bales.

Table 7-5 The Costs of a Perennial Resin Crop (Calotropis procera) at the Ord River Irrigation Area

Variable costs	$/ha
1. Seed $12/ha (equivalent annual cost)	2
2. Fertilizer	
50kg/ha P; 216kg Dap @ $408/t	88
15kg/ha N; 32kg urea @ $361/t	12
3. Water 8M litres @ $2.50/M litres	20
4. Insecticide	-
5. Fuel, oil and grease	25
6. Repairs and maintenance	25
7. Harvesting @ $20/t	400
8. Transport 3km @ 20¢/t.km x 20t/ha	12
	584
9. Fixed costs	
Management and labour	141
Miscellaneous	22
Machinery depreciation	94
Return on capital including land	113
Total Costs	954
Cost of resin crop delivered ($/t)(20t(DW)/ha.yr)	48

The total delivered cost at $48/t(DW) is comparatively low and 40% of the total cost is fixed costs. There might be some possibility of decreasing fixed costs per hectare by increasing farm size (Rawlins 1981).

7.6 The Costs of Producing Guayule in an Environment with Summer and Winter Rainfall

Cornforth et al. (1980) estimated the costs of guayule production in the USA at two sites, under irrigation and dryland conditions, assuming establishment from seed and seedling. The crop is harvested after 3 years and dug out after 5 years. Yields are much lower than we have estimated for Australia and we have calculated from Cornforth's results delivered costs of $144 to $184/t(DW) (US$1979). Cornforth's results show a saving in total costs of approximately 10% from seed establishment when compared with seedling establishment.

Foster et al. (1980) have assessed guayule technology and calculated an agricultural transfer price of US$(1979?)154 to 224/t(DW). This group states that yields of dry matter from varieties available in 1979 range from 2.7t/ha.yr for the dryland case to 5.6t/ha.yr for irrigated guayule.

Table 7-6 is a preliminary budget for a five year guayule crop grown in Tara Shire which is approximately 350km due west of Brisbane. The crop is assumed to be planted from seed, harvested with a forage harvester after three years and dug from the ground for a second harvest after a further two years. The yield assumed is 5t(DW)/ha.yr plus an extra 5t(DW)/ha of roots at the second harvest. The country currently produces average wheat crops of 1t/ha. The farming system is developed from Cornforth et al. (1980) and the costs are from Thew (1980). Transport costs are assessed at

Table 7-6 The costs of production of guayule at Tara Shire, approximately 350km west of Brisbane, Queensland

		$/ha
Year 1	Seed	12
	Fertiliser 40kgN/ha	16
	Herbicide	29
	Fuel, oil and repairs	24
		81
Year 2	Fuel, oil and repairs (spraying)	1
	Herbicide	23
	Gross margin on wheat foregone	37
		61
Year 3	Gross margin on wheat foregone	37
Year 4	Harvesting cost 30t/ha (50% M.C.) @ $10/t	300
	Transport to processing plant 30t/ha, 20km x 10¢/t.km	60
	Fuel, oil and repairs	6
	Fertiliser 40kgN	16
	Herbicide	23
	Gross margin on wheat foregone	37
		446
Year 5	Gross Margin on wheat foregone	37
Year 6	Harvesting cost 30t/ha @ $12/t	360
	Transport to processing plant 30t/ha, 20km x 10¢/t.km	60
	Gross margin on wheat foregone	37
		457

10¢/t.km as the crop is transported wet (50% moisture content) and will be denser in the transporting container.

The 'present value' (Merrett and Sykes 1973) of the stream of guayule costs at the 13% discount rate is given by:
$81 + 61/1.13 + 37/(1.13)^2 + 446/(1.13)^3 + 37/(1.13)^4 + 457/(1.13)^5 = 743$

The present value of receipts assuming a price of \$P/t(DW) is:
$15P/(1.13)^3 + 15P/(1.13)^5 = 18.5P$

The break even price equates the present value of costs and receipts, i.e. is the solution of 18.5P = \$743; P = \$40/t(DW).

Guayule, as a five year perennial, might be replacing more than wheat in the cropping system since the latter is only a winter crop and would be associated with grazing or alternative crops in the complete rotation. Below is an estimate of the fixed costs which might be associated with guayule cropping.

	\$/ha
Depreciation of fixed assets	13
Interest on fixed assets, excluding land	16
Miscellaneous overheads	7
Labour (1 man @ \$15 000 p.a., 500ha)	30
Interest on land capital, \$150/ha @ 13% p.a.	20
	86

If these were substituted in Table 7-6 to replace the gross margin on wheat foregone then the cost of guayule production would rise by \$9/t(DW).

The cost of crop establishment and the cost of harvesting are perhaps the weakest cost items in Table 7-6. Cornforth et al. (1980) assumed seedling costs of US(1979)\$258/ha which compares with our seed cost of A\$(1980)\$12/ha. If we assume a seedling cost of A\$(1980)\$258/ha then the cost of guayule rises by \$13/t. The harvesting cost at \$10/t(50% moisture) is based on an assumption of forage harvesting into a bin for direct transport to the processing plant. If this cost were to double then total costs would increase by \$22/t(DW).

At yields of 3t(DW)/ha.yr and 2t(DW)/ha.yr the basic cost of production would rise to \$49 and \$62/t(DW) respectively. If all the adverse cost movements were to occur simultaneously yields of 2t(DW)/ha.yr, fixed costs instead of gross margins, seedling costs of \$258/ha, and harvesting costs doubled, then the delivered cost would be \$140/t(DW).

The range of costs we have generated is from \$40 to \$140/t(DW) and our own subjective estimate of what would be most appropriate carried forward in future processing budgets is \$60/t(DW). The range of costs are below the range of the US estimates reported above primarily because irrigation was not used and other costs per hectare were lower.

7.7 Overview

The costs of delivered resin crops are summarised below:

Table 7-7 Summary table of crop production costs

Crop and site	Yield t(DW)ha/yr	Cost (delivered year round) $/t(DW)
1. Euphorbia lathyris in southern winter rainfall region	3	59
2. Euphorbia lathyris irrigated at Griffith, N.S.W	10	51
3. Asclepias rotundifolia in the eastern region of year-round rainfall	7.5	58
4. Calotropis procera in the northern region of summer rainfall	10	46
5. Calotropis procera at the Ord River, W.A.	20	48
6. Guayule at Tara Shire, Qld	5	60

The costs of production are very similar. This is not altogether surprising as all crops carry the basic harvesting cost of \$20/t(DW) and generally the extra resin crop yields of some areas are offset by the higher costs of those regions.

Taking the winter rainfall annual crop as an example, costs can be partitioned as:

Costs associated with growing the resin crop	\$42/ha
Opportunity cost of wheat crop foregone	\$66/ha
Harvesting cost	\$20/t
Transport cost	20¢/t.km

A function relating delivered cost to yield (y in t(DW)/ha) for a plant requiring 300t/day for 300 days/yr can be derived as;

Delivered cost (\$/t(DW)) = 42/y + 66/y + 20 + 5.8/y (see Appendix)

The graph of the equation is shown in Fig. 7-1. Note that it assumes that the yield of the resin crop changes without affecting the gross margin of the crop foregone. Table 7-8 illustrates the partitioning of total cost as yield changes and shows how sensitive costs per tonne are to low yields. Most of the sensitivity derives from the assumption that the gross margin foregone would be unaffected by the condition which brought about the low yield in resin crops and this might not be a plausible assumption. In the case of drought, for example, the yield of grain is likely to be more variable than the yield of total dry matter from the resin crop.

Table 7-8 Partitioning of delivered costs as a function of yield*
((\$/t(DW))

Yield t/ha	Growing	Harvest	Transport	Total
1	108	20	6	134
3	36	20	3	59
5	21	20	3	44

* Assuming a supply to a 300t/day processing plant operating for 300 days per year annual crop

The second feature in Table 7-8 of note is the minor part of transport costs. Appendix 1 describes the level of uncertainty in these costs. If process plant size increased to 1255t/day then average weighted haul would increase from 17km to 34km and transport costs of Table 7-8 would double.

In this analysis we have not examined the likely costs of a development of resin producing crops in country with very low rainfall and not examined the likely costs of a plantation type of development.

FIGURE 7-1

Costs of Euphorbia lathyrus in the southern winter rainfall zone as a function of yield per hectare.

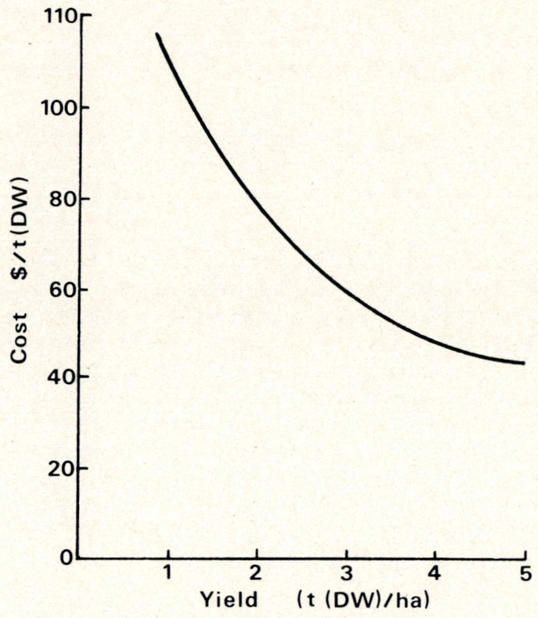

8 PROCESSING OF CROP MATERIALS AND RESIN COSTS

The first part of this section deals with solvent extraction technologies for separating the dried biomass from resin-containing crops into its various components, and the returns that might be obtained from those components. As assessment of resin cost was our major objective, we have identified likely returns from co-products, lignocellulosic bagasse, sugar, guayule leaf and rubber, and attributed all other costs to the resins. The nearest equivalent to plant resins are seed oils which have similar energy contents per unit of weight (40MJ/kg), and probably need less processing than would resins before use as a liquid fuel for transport. The assessment of seed oils as potential diesel fuels concluded that linseed, rapeseed and sunflower oils could be produced in medium sized plants for 38, 41 and 51¢/litre respectively in 1978-79 dollars (Stewart et al. 1981). At 1080 litres/t, the cost of rapeseed oil was \$443/t. Inflated 20% to 1980-81, the cost would be \$530/t = \$77/barrel o.e.. The processing of one tonne of rapeseed results in the production of 0.57t of oilseed meal. There are large world markets for these meals at \$112/t in 1978-79 dollars (= \$134/t in 1980-81 dollars). Fig. 8-2 shows the relationships of cost of feedstock and proportion of extractable resins in the feedstock compared with that needed to compete with seed oils.

The final part of this section deals with the technologies and costs of producing resins as a by-product to the much more valuable rubber from guayule. The most promising technologies for recovery of resins and rubber from guayule are solvent extractions similar to those described above. However, because of the previous commercial production and considerable research in USA and Mexico, we are able to report published methods and costs. Not all aspects have yet been proved in pilot or commercial application.

8.1 Technology for Extraction of Resins

On present knowledge it appears that solvent extraction as practiced with oilseeds is the only technology that is suitable for extraction of resins from field-dried crop materials (Kohan and Wilhelm 1980). The steps for processing seeds containing less than 20% oil (Australian Oilseed Crushers Association 1978) appears most likely to be appropriate for the processing of resin-containing crop material (see Fig. 8-1a). Some probable variations for resin-containing materials are outlined below:

FIGURE 8-1

Flow diagrams for (a) extraction of resins only (b) extraction of resins and of sugar syrup.

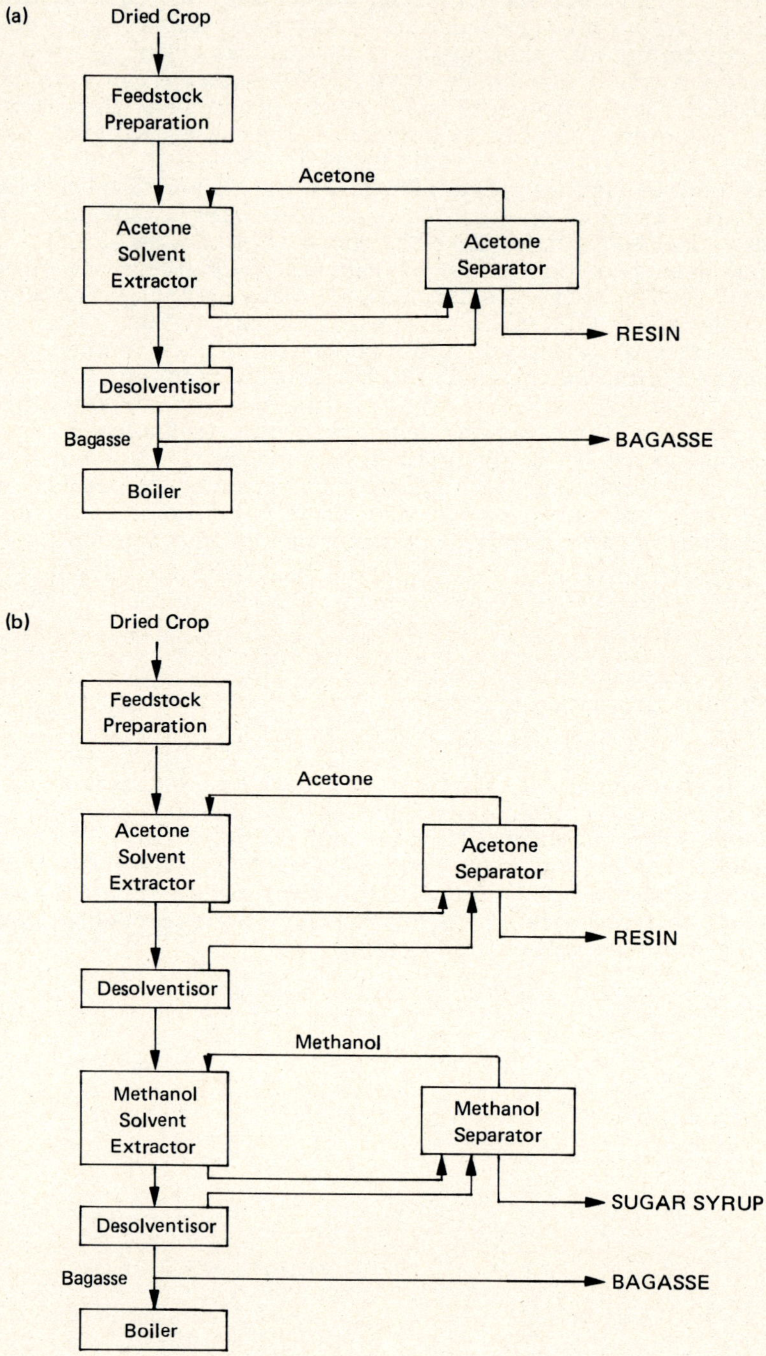

(a) Feedstock preparation. In low oil content seeds such as soybean, the feedstock preparation included removal of trash and foreign seeds on screens, cracking of seeds, cooking to coagulate protein and reduce moisture content to a low level to ensure efficient oil extraction, removal of hulls by aspiration, and flaking of seed material through rollers so that the solvent has ready access to the oils and so that the solvent is afforded reasonable percolation through the seed mass. In Euphorbia lathyris only about 10% of the resins are in the latex (personal communication: E. Nemethy). The heptane soluble fraction is mainly distributed in the leaves with much lower content levels in the stems. Thus it appears likely that resin-containing crop material will need to be ground to expose resins to the solvent, cooked to coagulate protein and to reduce moisture content to the optimum level, and pressed into flakes or pellets to permit reasonable percolation of the solvent. The energy cost of grinding plant material is directly related to moisture content of the crop material, because it becomes more brittle at lower moisture contents. Thus it may be desirable to cook coarsely ground or chopped material before grinding to small sized particles.

(b) Solvent extraction. Several types of solvent extractors are available, but they all operate on the same principle of repeated countercurrent washing with solvent-oil mixtures of increasing oil content. Oilseed materials entering the solvent extractor have oil contents of about 20%. For oilseeds with oil content of 30 to 50%, the flaking process as used for soybeans is replaced by a mechanical 'pre-press' which reduces oil content to about 20% as well as converting the seed to a suitable form for solvent extraction. The normal solvent is hexane. The oil content of the residue is normally reduced to about 1%.

For resin containing crops we assume that acetone can be used in commercial solvent extractors to remove resins I and II, leaving 1% resin in extracted material.

(c) Solvent-oil separation. The hexane-oil mixture is passed through an evaporator to remove most of the solvent, and the remainder is flashed off in a stripping column with the aid of steam and/or a vacuum. The same equipment should be suitable for acetone solvent recovery.

(d) Solvent recovery from extracted material. This is carried out in a 'desolventisor' using live as well as indirect steam to flash off the solvent. The same equipment should be suitable for recovery of acetone from extracted crop material.

(e) Process steam. Steam is required for cooking of feedstock, and recovery of solvent from resin and from extracted material. For processing of oilseeds, biomass fuel (e.g. as hulls from sunflower) equivalent to 10% of the weight of seed is burnt in a boiler to provide process steam. For resin-containing crops we propose that extracted crop materials, equal to 10% by weight of feedstock, be burnt as fuel for generation of process steam.

(f) Electricity. Approximately 55kWh are required for the processing of one tonne of oilseeds. We assume that the same quantity will be required for one tonne of resin-containing crop material, and that it will be purchased at 3¢/kWh.

8.2 Cost of Extraction of Resins

In the absence of commercial data for extraction of resins from crop material, we assume that costs per tonne of feedstock are the same as costs per tonne for processing of oilseeds. Data obtained from the oilseed crushing industry by Stewart et al. (1981) have been updated to December 1980 dollars (see Table 8-1). For the large scale plant (1360t/day) that operated on dehulled sunflower, the capital costs were reduced by excluding dehulling equipment, and throughput was taken as the tonnage throughput of dehulled sunflower seed = 1225t/day. For the 300t/day plant, the reported cost is that for rapeseed and linseed. In subsequent calculations the costs are rounded to $36 and $18 per tonne feedstock.

Table 8-1 Estimated costs of resin extraction in two sizes of plants (1980-81 Australian dollars), excluding feedstock cost

	300t/d plant = 90 000t/yr		1225t/d plant = 367 500t/yr	
Fixed capital	$5.45 million		$14.2 million	
Annual costs	$000	$/t feedstock	$000	$/t feedstock
operating costs	2185	24.3	3741	10.2
capital charges (20%)	1090	12.1	2840	7.7
Total	3275	36.4	6581	17.9

8.3 Potential Use and Value of Co-product Bagasse

Processing of resin-containing crops would result in large amounts of lignocellulosic bagasse for which there is no market at present. Data from Wan et al. (1979) was used to estimate the cost of converting that residue to methanol in a plant processing 908t(DW)/day (see Table 8-2). The size

Table 8-2 Estimates of the cost (1980-81 Australian dollars) of converting lignocellulose to methanol, derived from Wan et al. (1979) excluding feedstock cost

Total methanol production	128 850 t/yr
Operations	330 days/yr
Feedstock rate	2.33t(DW)/t methanol
Fixed capital	$117 million
Annual costs $/t methanol	
operating	52
capital charges (20%)	182
TOTAL	234

selected would utilise all the lignocellulose residues from a large scale resin plant, or from one medium scale plant plus three times that amount of residues from other crops. If the cost of the methanol per unit of energy was the same as that for the seed oils from medium sized plants, the bagasse feedstock would cost $27/t(DW). That cost appears appropriate in relation to other residues, e.g. cereal straw.

8.4 Cost of Resins

The materials balance and estimated cost of resins in medium scale plants is shown in Table 8-3. The cost of resins was calculated as follows:

Table 8-3 Materials balance and costs of resins in medium plant (1980-81 dollars) for various species in various regions and energy value of resins and methanol products

	Euphorbia lathyris southern	*Euphorbia lathyris* southern irrigated	*Asclepias rotundifolia* eastern	*Calotropis procera* northern irrigated	*Calotropis procera* northern	Rapeseed dryland Dubbo NSW
I MATERIALS BALANCE						
Proportion of resins	.08*	.08	.10	.055	.055	
extractable resins	.07	.07	.09	.045	.045	.40 (oil)
fuel	.10	.10	.10	.10	.10	
losses	.01	.01	.01	.01	.01	.03
bagasse residue	.82	.82	.80	.845	.845	.57 (meal)
II COST OF RESIN - MEDIUM PLANT						
(a) Sale of resin only						
Cost of feedstock $/t	57	51	58	46	48	217
resin extraction $/t feedstock	36	36	36	36	36	66 **
TOTAL $/t feedstock	93	87	94	82	84	
Resin cost $/barreloil equivalent***	192	180	151	264	270	
(b) Sale of resin + bagasse						
By-product return from bagasse $27/t	22	22	22	23	23	76 (meal)
Total $/t feedstock	71	65	72	59	61	207
Resin cost $/barrel oil equivalent	147	134	116	190	196	75
III ENERGY VALUE OF PRODUCTS GJ/t feedstock						
Resin	2.6	2.6	3.4	1.7	1.7	16.0 (oil)
Methanol from bagasse	7.1	7.1	7.0	7.4	7.4	

* resin I plus resin II for *Euphorbia lathyris* from Nemethy *et al.* (1980), resin I + hydrocarbons for others from Table 3-4 ** total oil extraction includes $36 for oil extraction, $11 for working capital, $17 for storage of seed and product, and $2 for fuel (crop residues) *** energy content of one barrel of crude oil = 5.8GJ. Assuming resins have energy content of 40GJ/t then one barrel o.e. = 0.145t resin, or 6.9 barrel o.e. per tonne resin.

let f = cost of feedstock in \$/t(DW)
b = value of bagasse residue in \$/t(DW) of bagasse
p = cost of processing in \$/t(DW) of feedstock
r = cost of resin in \$/t(DW) of resin
R = proportion of recovered resin
B = proportion of surplus bagasse residue
then $r = (f + p - bB)/R$

As shown in Table 8-3, bagasse equivalent to 0.1 crop dry weight is used as fuel, and 0.01 is lost in processing. Thus $B=1.0-R-0.11$, and $r=(f+p-b(1.0-R-0.11))/R$.

When resins only are sold the cost of resins range from \$151 to \$264/barrel o.e., far above seed oils as represented by rapeseed oil also shown in Table 8-3. The returns from sale of bagasse residues are about two thirds of the cost of resin processing, and resins then range from \$116 to \$196/ barrel o.e. The resin cost is very sensitive to feedstock costs - for each \$1 rise in the cost of feedstock, the cost of resin rises by the reciprocal of the proportion of extractable resin. The cost in large sized processing plants is decreased by \$18/t feedstock, which decreases resin cost by 18 times the reciprocal of the proportion of extractable resin.

Data on the resin-containing crops are shown in a different manner in Fig. 8-2. Values for recoverable resin content and feedstock cost for the candidate crops are shown together with lines that indicate the levels that would be needed to break even with oils from oilseeds. To compete by sale of resin only with existing achievements of seed oils in medium plants selling oil plus meal the recoverable resin content would need to be increased to 0.15 to 0.20. That level might be achievable by growing resin containing crops to physiological maturity, or by growing some Eucalyptus spp.

Fig. 8-2 also shows that, if only the resin is sold and if the feedstock is free, the proportion of extractable resin would need to exceed 0.069 to justify the cost of extraction. Where both resin and bagasse are sold the minimum proportion of resin required to justify extraction would be 0.079 for medium plants. In the latter case it would not be worth extracting resin from Calotropis procera or Euphorbia lathyris.

The development of a market for the bagasse from resin crops by conversion to methanol would call for investment of over \$100 million, about eight times the capital required for resin extraction (see Tables 8-1 and 2). The enterprise would then become primarily methanol from biomass, as the energy value of the methanol from bagasse would be two to four times the energy value of resin (see Table 8-3). Domesticating and developing new crop species for methanol production is not needed because there are 17Mt(DW) of existing lignocellulosic residues in Australia potentially available for conversion to methanol. However if recoverable resin content could be increased to 0.17 the energy contents of the resin and the methanol products would be equal. Setting up a methanol plant for conversion of bagasse from a medium plant would not be justifiable, but if conversion of crop residues to methanol was undertaken, it would improve the economics of production of resin-containing crops. In setting a target level for resin content for competition with seed oils we assume that resins only will be sold, which will require resin contents of 0.15 to 0.20.

FIGURE 8-2

Proportion of extractable resins and feedstock costs for candidate crops in relation to those needed to break even with seed oils, for sales of resin only and resin plus bagasse from a medium sized plant.

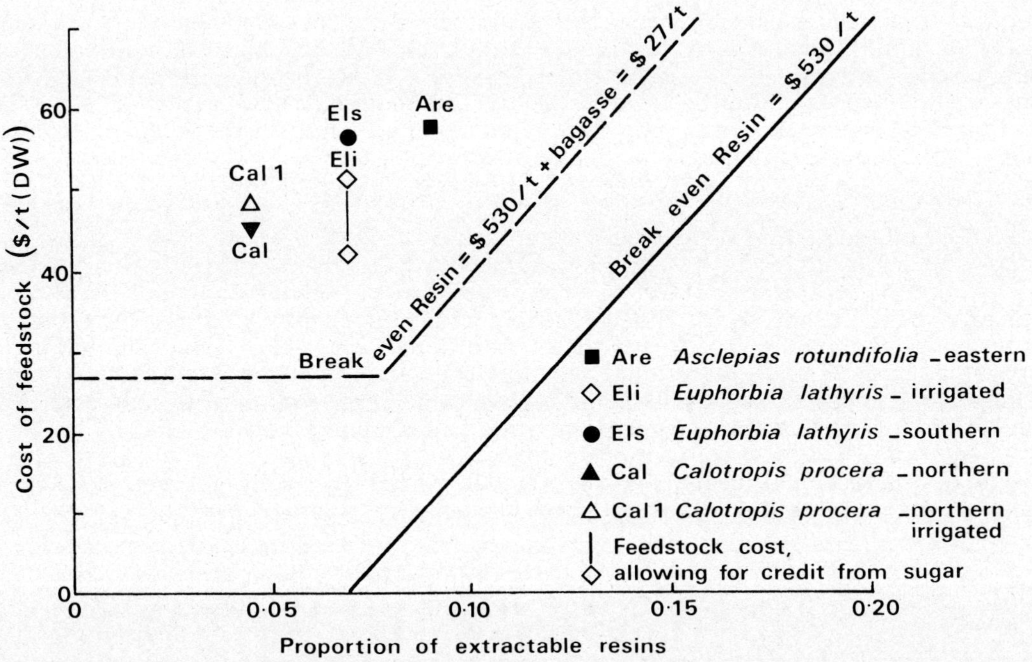

8.5 Technology for Recovery of Sugars

Nemethy et al. (1980) reported that the water-soluble fraction of the methanol extract of the residue from heptane made up 27% of the dry weight of the crop, 20% being simple sugars, 4% amino acids, and 3% of unknowns, probably flavanoids.

Kohan and Wilhelm (1980) examined the possibility of recovering the sugars by washing in water. This approach has two main drawbacks. Firstly, the sugars are recovered as a dilute 10% solution, which limits their use to immediate fermentation to ethanol. Secondly the lignocellulose has a very high affinity for water, and the proposed dewatering of the residue involves expensive rotary drum filtering and results in a final product that contains 50% water (wet basis), which lowers its value appreciably.

We assume that the sugars could be recovered in a solvent extraction plant similar to that used for resins, but using methanol as the solvent. In recovery of the methanol from the dissolved material, we assume that steam addition would be used to produce a syrup of 70% dry matter, equivalent to 52% simple sugars, i.e. similar to molasses. This second solvent extraction set would process the residue from the acetone extraction plant (see Fig. 8-1b). We assume a steam use equivalent to half that of the resin extraction plant, i.e. equivalent to 5% of feedstock crop material. We also assume that 90% of the sugar will be recovered in the commercial extractor.

8.6 Cost of Extraction of Sugars

Additional capital would be required for a second solvent extractor, solvent-sugar separator, and solvent-residue separator, plus the cost of increasing the boiler size by 50%. Additional operating costs include two operators per shift, additional electricity, and cost of methanol solvent. These total costs, summarised in Table 8-4, are rounded to $10 and $5/t feedstock for medium and large processing units respectively.

Table 8-4 Additional costs for the separation of sugars by solvent extraction with methanol in two sizes of plant

	Crop material input			
	300t/d plant = 90 000t/yr		1225t/d plant = 367 500t/yr	
Fixed capital	$2.4 million		$4.3 million	
Annual costs	$000	$/t feedstock	$000	$/t feedstock
operating costs	570	6.3	910	2.6
capital returns (20%)	350	3.9	930	2.4
TOTAL	920	10.2	1840	5.0

What market price might the sugars command in the form of a syrup like molasses with 70% dry matter and 52% sugars? Sugarcane molasses with approximately 50% sugars cost $40/t ex sugarmill = $80/t of sugars. Grain sorghum and barley both contain about 11% crude protein and 62% starch (equivalent to 65% sucrose). Taking 33¢/kg of crude protein as in rapeseed in 1978-79, inflation of 20% would give $44 for the crude protein in one tonne of grain. A price of $105/t grain in rural centres would thus give $61 for

650kg of sucrose equivalent = $94/t of sucrose equivalent. We arbitrarily take a value of $100/t sugars for the following calculation.

For Euphorbia lathyris with 18% extractable sugars, the value of sugar extracted is $18, giving a margin of $8 and $13/t of feedstock for medium and large extraction plants when only resin is sold. However, when bagasse residues are also sold, the margin is reduced by $8 because of the lower return from bagasse residues (see Table 8-5). Assuming that sugar and other methanol solubles are always in the ratio of 3:1, the returns per tonne of feedstock for extraction of sugars are as shown in Fig. 8-3. Returns from sugars exceed extraction costs only when the proportion of extractable sugars exceed .05 for large plants with resin sales only, .10 for medium plants with resin sales and large plants with resin plus bagasse sales, and .18 for medium plants with resin plus bagasse sales.

FIGURE 8-3

Relationship between proportion of extractable sugars in feedstock and margin over total costs of extraction of sugars for large and medium sized plants with or without sales of bagasse.

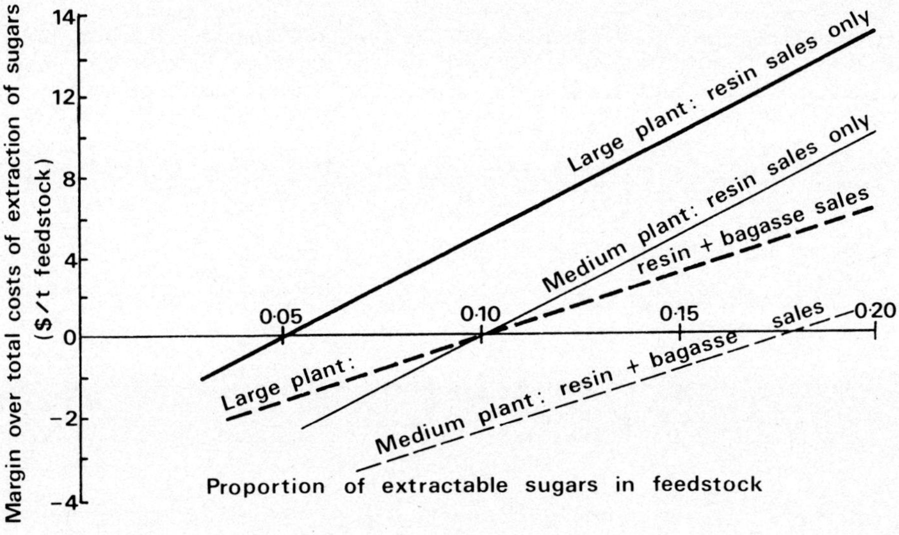

The effect of a sugar by-product credit of $8 for medium plants with resin only sales can be allowed for in Fig. 8-2. For example Euphorbia lathyris grown under irrigation has a feedstock cost of $51, and allowance of $8 for sugar credits for a medium plant would reduce that feedstock cost to $43. This improvement is small compared with that needed to be competitive with seed oils.

Table 8-5 Material balance per tonne (DW) of feedstock for sugars extracted from *Euphorbia lathyris*

	Dried crop t	Acetone extract t	Acetone residue t	Methano extract t	Methanol residue t
Resin	.08	.07	.01		.01
Sugar	.20		.20	.18	.02
Other methanol solubles	.07		.07	.06	.01
Remainder	.65		.64		.63
Loss			(.01)		(.01)
Total	1.00	.07	.92	.24	.67
less fuel			(.10)		(.15)
Bagasse residue available for sale			.82		.52

8.7 Technology for Extracting Rubber and Resins from Guayule

The rubber in guayule is not contained in ducts but as a latex in soft-walled cells distributed throughout the stems, branches and roots. There is no rubber in the leaves. Resin occurs in ducts found throughout the shrub (National Academy of Sciences 1977).

The technology used in Mexico in the period 1905 to 1951 is shown in a schematic diagram in Fig. 8-4, based on data from Velazquez (1978) and Taylor (1978). Stored guayule was used, but subsequent research in USA (Taylor 1978) and Mexico (Campos-Lopez *et al*. 1978) showed that rubber yield was lowered and molecular weight of the rubber was reduced with storage. Defoliation was not normally practiced, but the shrub was soaked before milling. Grinding was normally carried out in two stages - a coarse milling or grinding followed by a pebble mill. The slurry from the mills was fed into large tanks in which the rubber crumbs, cork, and some plant debris floated to the surface and was skimmed off. The bagasse sank to the bottom and was recovered, dried and used as boiler fuel. The mixture of rubber, cork and debris was subject to low pressure, which forced the air out of the cork and debris and allowed them to sink in the secondary flotation tank. The rubber was then dried and baled. Rubber produced in this way contained high levels of impurities, particularly resin.

FIGURE 8-4

Flow diagram for extraction of rubber from guayule as used in Mexico 1903-1951 (adapted from Velasquez et al. 1978).

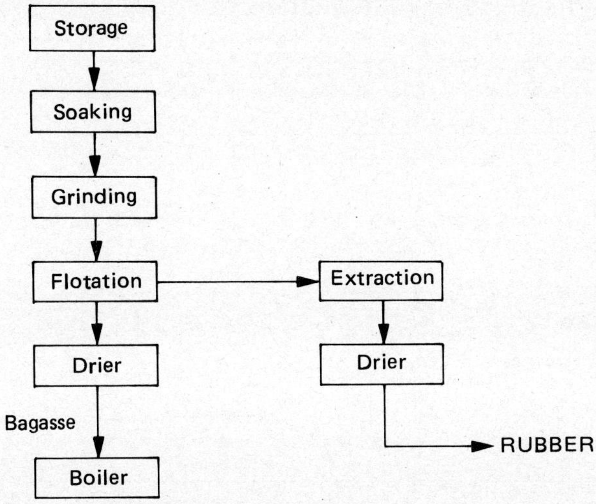

During the Emergency Rubber Project in USA the guayule was defoliated by parboiling and screening, and trials were carried out with wood pulping equipment, e.g. Jordan mills , for the final grinding operation (Taylor 1968).

The technology of the Saltillo pilot plant in Mexico (Campos-Lopez et al. 1978) is illustrated in Fig. 8-5. Freshly harvested shrub was used, and Bauer mills, as used in the wood pulp industry, were used for the second grinding stage. One alternative involved bagasse separation by flotation before deresination by solvent extraction with acetone and gave resin yields equivalent to about 3% of crop dry weight. Another alterative had resin extraction before bagasse separation and gave a resin yield of about 10%. In both cases only the crumb rubber with cork and debris went on to solvent extraction of rubber with hexane. The rubber was coagulated by adding water and the hexane was recovered for reuse. Experience with the pilot plant showed that it was possible to produce rubber of high purity, similar to the better *Hevea* products.

FIGURE 8-5

Flow diagram of the Mexican Guayule pilot plant that commenced operations at Saltillo in 1976 (adapted from Campos-Lopez et al. 1978).

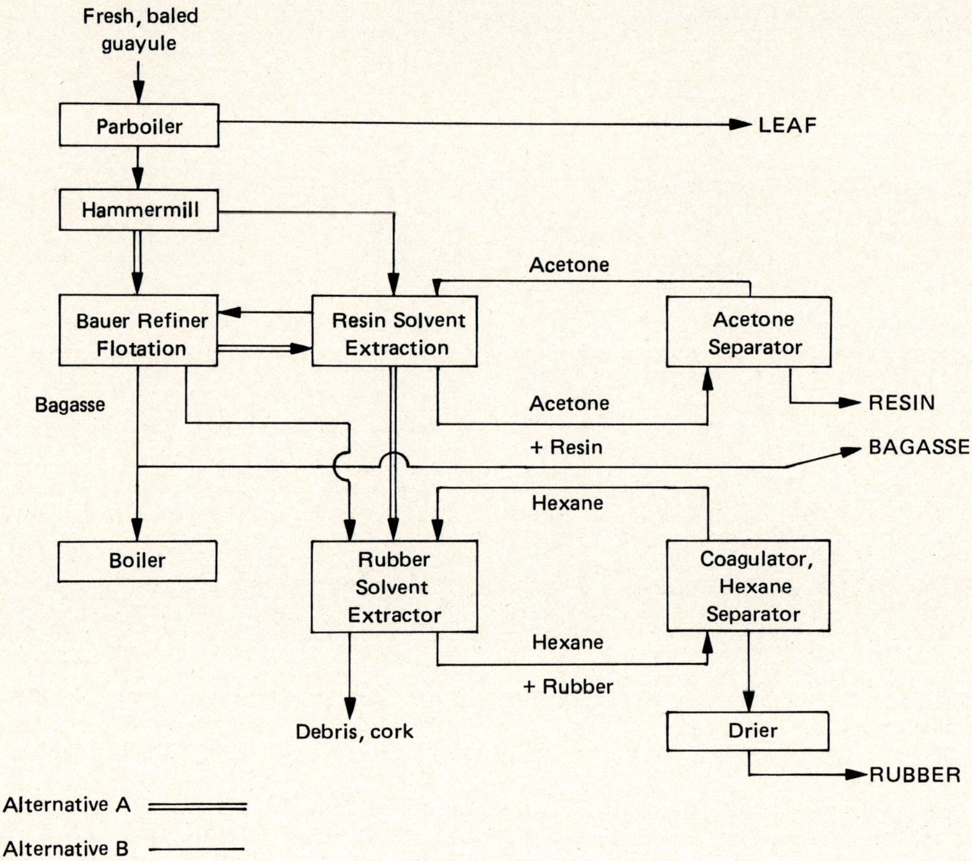

A conceptual plan for production of rubber from guayule in plants processing over 100 000t/yr of guayule was put forward by Nivert et al. (1978), and is illustrated in Fig. 8-6. This does not employ flotation for separation of bagasse, but passes all of the crop material through the resin and rubber solvent extractors. The acetone solvent serves the two functions of dissolving the resins and dehydrating the crop materials. The desolventisors to recover solvent are the same as those used in the oilseed industry. The plan is yet to be tested in practice, but it should give high yields of resin and high quality rubber.

FIGURE 8-6

Flow diagram for solvent extraction of resin and rubber from guayule (adapted from Nivert et al. 1978)

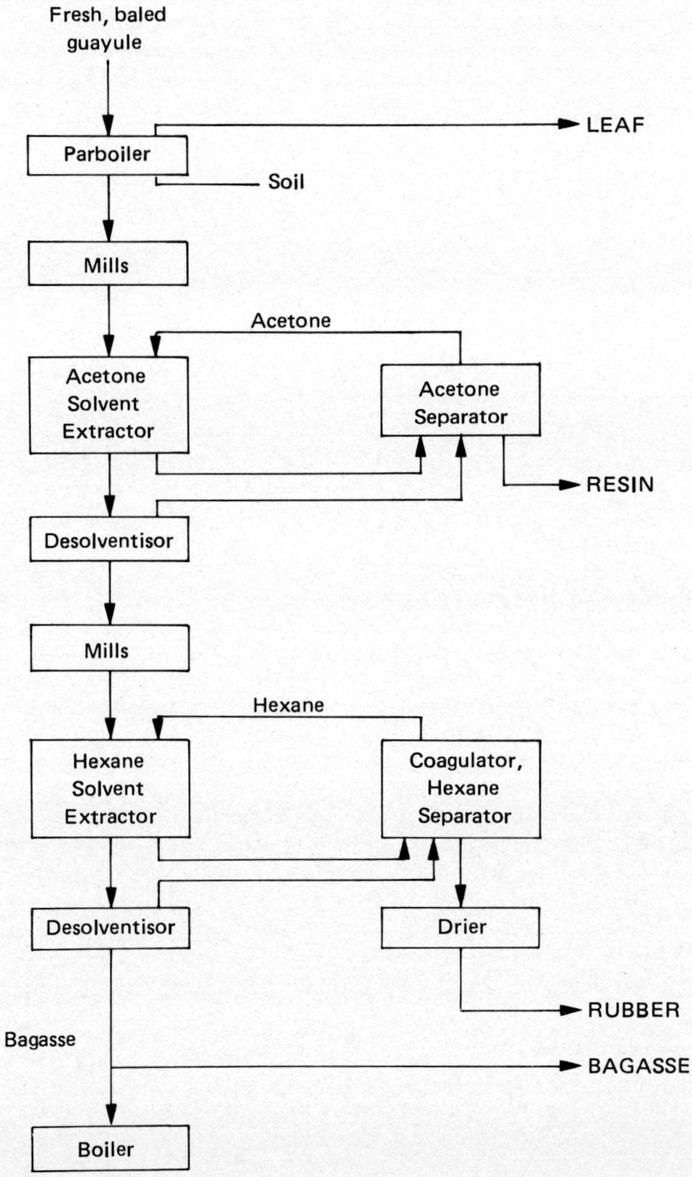

8.8 Cost of Extraction of Rubber and Resin from Guayule

The only published estimates of costs of extraction of rubber and resin in plants of sizes comparable to those reported earlier is from Nivert et al. (1978), and unpublished data of Nivert as reported by Foster et al. (1980). The published data was for 25 million lbs/yr rubber, equivalent to 56 750t/yr feedstock, but the costs were given in 1985 US dollars. The unpublished data were for a plant of 50 million lbs/yr rubber, equivalent to 113 500t/yr feedstock. The capital cost of $18 to $20 million and extraction costs of 20¢/lb rubber were assumed to be in 1979 US dollars, which were assumed to be the same as 1980-81 Australian dollars.

Table 8-6 Estimated cost of rubber and resin extraction from guayule (1980-81 Australian dollars)

Throughput of feedstock	113 500t/yr = 344t/day	
Fixed capital	$20 million	
Annual costs	$000	$/t feedstock
Operating)	)	
Storage-products)	6000)	53
Working capital)	)	
Capital return (20%)	4 000	35
Total	10 000	88

The capital and annual operating costs are higher than the combined resin plus sugar extraction in the large plant of 1225t/d, while the throughput of this plant is 344t feedstock/d. As a result the cost of processing the guayule is nearly twice the cost of processing *Euphorbia lathyris* for resin and sugar.

8.9 Cost of Rubber and Resins from Guayule

The materials balance for processing guayule are shown in Table 8-7. Losses were not listed by Nivert et al. (1978) but were assumed to be the same as for the processing of oilseeds. Bagasse used as fuel was arrived at by difference. It is a much higher proportion of feedstock than that proposed for resins (.10) or resins plus sugars (.15) from other crops.

The cost of production of rubber and resins is shown in Table 8-8.. Two figures are used as cost of feedstock - they are the middle of the range in our own assessment (see Section 7.6) and the highest reported by Cornforth et al. (1980) from Texas.

Table 8-7 Materials balance for extraction of rubber and resin by solvent extraction from guayule (Nivert et al. 1978)

Extractable rubber	.20
" resin	.10
Leaves	.15
Surplus bagasse	.16
Losses	.03
Bagasse used as fuel	.36

Table 8-8 Cost of production of rubber and resins from guayule in medium sized plant

	This report	Cornforth *et al.* 1980	
Cost of one tonne feedstock	60	184	Cost
of processing one tonne feedstock	88	88	
	148	272	
Returns from .20t rubber @ $950/t	190	190	
.15t leaves @ $40/t	6	6	
.16t surplus bagasse @ $27/t	4	4	Cost
of .10t of resin	-52	72	Cost
of barrel of oil equivalent	-79	104	

The price of natural and synthetic rubber in relation to OPEC crude oil is shown in Fig. 8-7. The similarity in these trends is clear. The minimum OPEC price, that of Saudi Arabia, is now $US32/barrel, the Financial Review recently (10/8/81) reported US price of 55¢/lb for rubber. For Table 8-8 a cost of 50¢US/lb for rubber (= 44¢A/1b = A$950/t) is taken. Bagasse has been costed at $27/t as earlier. Guayule leaf, which has a crude protein content of 7 to 15% (Gomez-Contreras 1978) has been given a nominal value of $40/t as feed for ruminant animals. For our estimated feedstock cost returns from other products exceeds the total costs by $79/barrel o.e. Fig. 8-8 shows that feedstock cost in the range of $40 to $140/t(DW) are associated with resin costs in the range -$100 to +$40/barrel o.e. For the Texas price of raw materials, the cost of resin is $104/ barrel o.e., which is higher than seed oils.

FIGURE 8-7

Relationship between the cost of natural and synthetic rubber and OPEC crude oil prices (Foster et al., 1980)

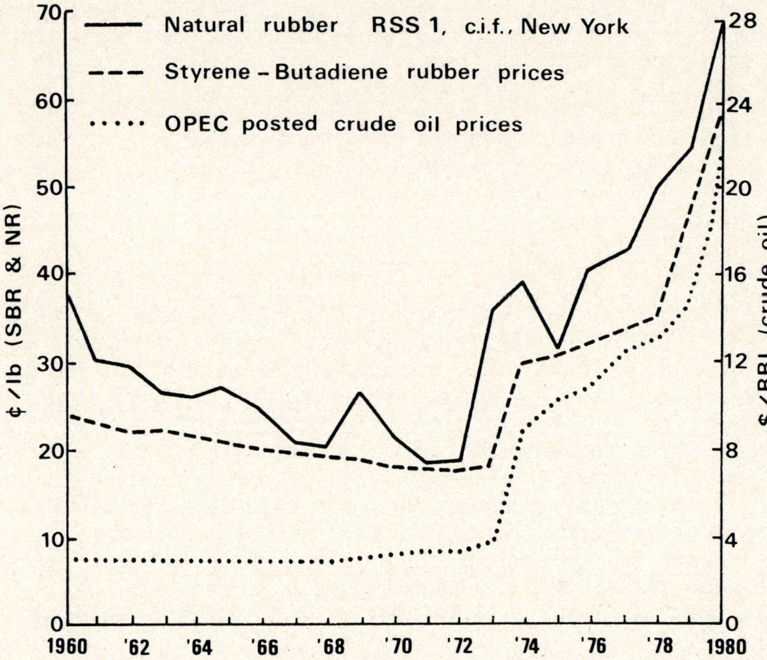

FIGURE 8-8

Relationship between the cost of guayule and the cost of resins (1980-81 dollars)

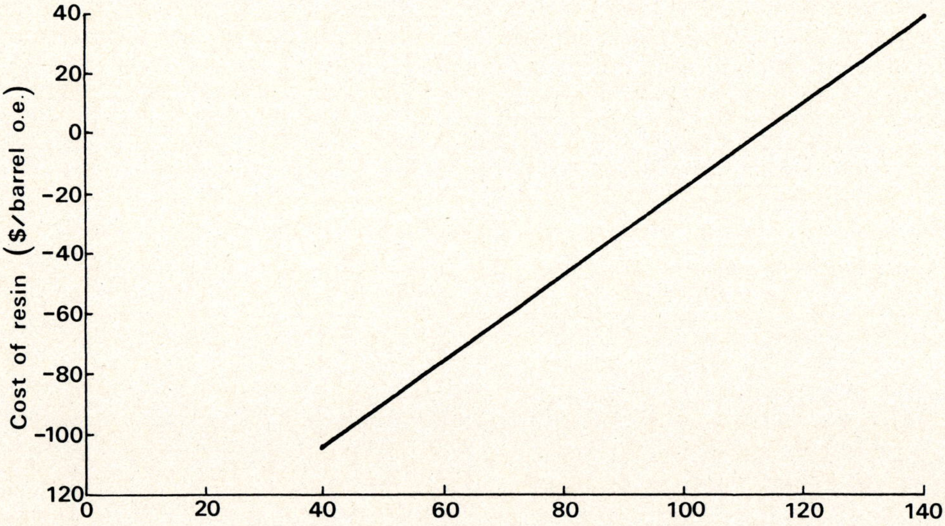

9 DISCUSSION

This assessment has been very tentative because little direct data were available and it was necessary to rely on indirect data. However, the exercise has been worthwhile in that it identifies weak points in the proposed production systems which are commented on below, and also arrives at conclusions for future research which are reported in the following section.

9.1 Limitations in Crop Production

The major weakness is the lack of information on dry matter yield and content of resins, and further research on these aspects is essential.

Our suggested technologies for harvest, storage, and transport are not fully adequate. They assume that crops of *Euphorbia lathyris*, *Asclepias rotundifolia* and *Calotropis procera* would be mown and conditioned, field dried, gathered into large round bales, stored in the field with no loss for periods of four to nine months, and transported by road to the processing plant on a year round basis. Storage losses are likely to be very high for sugars in *E. lathyris*. However those losses might be avoided if the crop is grown to physiological maturity and those sugars converted to resins in the seed. Peoples *et al*. (1981) reported that resins persisted in senescent material, and they probably would be much less prone to loss than the sugars. Conventional roofed storage is prohibitively expensive for low density bales, and low cost plastic covers, as for temporary wheat storage, are not practical for round bales. Transport of the low density round bales is also difficult. We do not recommend immediate research on these aspects as we are hopeful that future investigations of harvesting, storage, and transport of cereal crop residues will identify technologies that may also be practical for storage and transport of resin crop materials.

One aspect that has not been discussed is the variability of annual yield that is characteristic of our cropping areas. For example in the ten years 1967-68 to 1976-77 the average yield of wheat in N.S.W. was 1.25t/ha, varying from 0.71t/ha in 1967-68 to 1.65t/ha in 1976-77. Vegetative yields are less variable than grain yields, and extreme yields might possibly vary by ±20% around the mean. Long term storage of raw material is impractical, and no other alternative is apparent. This means that in bad years the processor would be forced to close his plant for part of the year and in years of peak yield the farmer would not be able to sell all that he had produced. Both would want a higher than average return on their investment money to cover those contingencies, for which we have made no allowance.

9.2 Limitations of Processing

The technology of solvent extraction of sugars with methanol is completely untried but we consider that research to assess that technology should not be undertaken until the growing of *E. lathyris* to maturity is tested. If our hypothesis is correct, there will not be any sugars to extract, and all of the energy rich components will be recovered by a single solvent extraction with acetone. Laboratory analysis (L.R. Williams: unpublished data) indicate that *E. lathyris* is unique in its high sugar content and that other candidate species have only low levels of sugar.

The acetone extraction of resins has not been carried out commercially as yet, but it is being used in the guayule pilot plant in Saltillo in Mexico, and in that application the acetone solvent extraction performs the dual functions of resin removal and water removal. Guayule is likely to reach commercial scale operation before any of the other crops that we have considered, and it should be possible to extrapolate from that experience to the needs of other crops.

9.3 Overall Limitations to Plant Resin Production Systems

Apart from guayule, in which the production of surplus bagasse is small and of little economic importance, all other plant resin production systems result in the production of large amounts of lignocellulose bagasse for which there is no current market. If crops can be developed with 20% of recoverable resins, the unused residue will still constitute two thirds of the biomass yield. Can the unique properties of the bagasse (dry, finely divided) be exploited for specific uses? For example, could the bagasse be conveyed and metered in a similar way to 'dried wood product' as a fuel in brick kilns (Kemp 1981). One medium sized plant for processing resin-containing crops would produce about eight times as much bagasse as might be used in one brick kiln. Are there other possible similar uses? We cannot answer these questions at the present time, but we will include this bagasse in our future studies on potential sources and uses of of lignocellulose as fuels.

We have not made provision for cost of storage of product and cost of working capital, but they would be very much smaller than these items for oilseeds because of the low cost of the raw material and the low content of high value product-resins.

The plant resins cannot be readily used directly as diesel fuels as they are dark coloured solids at ambient temperature and melt to viscous fluids at temperatures above ambient. It may be possible to use them directly by heating the fuel supply system as has been done experimentally with tallow, but we are not aware of any attempts to do that. The only data on Mobil catalytic conversion costs are for very large plants processing 25 000t coal/day to methanol-gasoline (Lee et al. 1980), in which the inferred cost is about $2 to $3/barrel oil equivalent. This can be taken as an absolute minimum cost. We do not know if it would be desirable to process plant resins separately, or whether they might be mixed before processing with other materials, e.g. crude oils from coal.

Kohan and Wilhelm (1980) reported that gum rosins and steam - distilled rosins (naval stores) from wood command prices of $1.50 to $0.60/kg, but there is no published data on the suitability of plant resins for those markets. Australia imported less than 10 000t of resins and associated products in 1978-79 so this alternative is not significant.

Taken overall, it does not appear that plant resins have the flexibility of alternative uses as do seed oils (for fuel or food), nor could they be processed and used as simply in the case of an emergency shortage of liquid fuel.

We have not identified any advantages that resin-containing crops might have over oilseed crops, and have in fact concluded that resin-containing crops should be treated more as seed crops in order to maximise resin content (which would include seed oils). Oilseed crops have the following characteristics: they concentrate all their oil in their seed (30 to 50% oil), only a small proportion of the crop is harvested, it is harvested when dry and is easy to transport and store, the oil can be removed from the seed at very low cost per unit of oil because of the high oil content, and the value of the by-product meal makes a very significant contribution to the returns from the crops. Oilseeds also produce lignocellulose which remains in the field. However it could be harvested and delivered for about the same cost as we have allowed for bagasse from resin-containing crops, i.e. $27/t(DW).

10 CONCLUSIONS

1. The economics of production of resin as a by-product of rubber from guayule appear very attractive. Dr. K. Henderson, an ANU research fellow funded by the CSIRO Division of Land Use Research, has visited USA and Mexico to collect further information on guayule. We will make a more comprehensive assessment of guayule later this year, which will take into account the interest and activities of the State Departments of Agriculture.

2. Several Eucalyptus spp. have contents of essential oils plus resins in the range of 0.12 to 0.20 of their dry weight. Further data are being sought with the objective of making economic assessment of the most promising species.

3. Other species have been reported to have resin contents that are much too low to provide resins at a cost competitive with seed oils. However two areas of research hold promise of possibly leading to higher resin contents that might be economically competitive:

(a) crops harvested near maturity appear to have appreciably higher resin contents, and short-term field experiments (see 4 below) could evaluate that possibility; and

(b) further chemical research is needed to ascertain if further components of the plants (acetone soluble-heptane insoluble) have high energy contents and whether they could be converted catalytically to liquid fuels. Appropriate analytical methods should then be identified for application to existing plant samples and those that arise from (a) above.

4. Crop yields have had to be estimated indirectly. Short-term, but comprehensive field experiments at two or three selected sites are essential to provide productivity data.

5. In the USA some thousands of species have been examined for resin and rubber content, but only two or three species have been identified with high resin content. Basic plant physiology - biochemistry has not provided useful leads on likely candidate species. Ether extract analysis as performed for animal feedstuffs appears to offer promise as an additional survey tool because of the very large number of species that have been analysed by that method. At the same time it appears desirable to have a small continuing program of collection of possible candidate species by competent biologists who are adequately informed on key aspects of sampling.

APPENDIX 1: DETAILS OF THE COST OF GROWING Euphorbia lathyris IN THE EASTERN WHEATBELT OF WESTERN AUSTRALIA

This appendix is included to illustrate the level of uncertainty associated with the budgets that have been prepared.

1. Seed — Mendel et al. (1979) reports that seed costs in the USA are US$(1978) 30/lb for seed produced for experimental purposes only. At an application rate of 2lbs/acre, seed costs are then high. Total costs of $US148/ha compare with current Australian wheat seed costs of $6/ha. In a situation of large scale production, costs would fall and a seed cost of $12/ha is included in this budget.

2. Fertilizer — Because of little evidence of nitrogen response in overseas studies of Euphorbia lathyris (Sachs et al. 1981) no nitrogen fertilizer is assumed in the budget (9kg/ha of nitrogen is added to wheat grown in the area). Phosphorus is used at the rate currently applied to wheat in the area, i.e. 100kg superphosphate/ha. At a cost of $100/t, that application costs $10/ha.

3. Sprays — Weeds were a serious problem in the experiments reported by Sachs et al. (1981) and Mendel et al. (1979) reports that all USA test plots have been weeded by hand. Herbicides used are assumed to be 1 litre/ha of Hoegrass @ $15/litre. No treatments have been assumed for pests and diseases. This is despite reports in the literature of significant aphid attack on experiments in the USA (Mendel et al. 1981) and diseases on non-irrigated crops (Sachs et al. (1981).

Tractor operation

The system of plough, cultivate, seed is as described by Poole (1980) and represents a suggested minimum 'conventional' practice. Minimum tillage is widely practised in Western Australia and it seems likely that hydrocarbon crops would be grown accordingly if suitable herbicides are discovered.

Rates of work are taken from Eckersley (1979) and costs per hour are updated from Eckersley (1979) using the BAE Index of prices paid by farmers.

	$/hr
Fuel $8.20/hr @ 21¢/litre updated to 30¢/litre	11.71
Repairs $6.00/hr updated by BAE spare parts index 466/385	7.26
Oils and greases @ $1.50/hr updated by fuels index 562/431	1.96
	20.93

The spraying costs are based on the use of a 45kWPTO tractor as costed by Bailey and Buffier (1981). Costs are $5.00/hr.

Harvesting

Harvesting is a most important component of total agricultural costs. As far as we know no experiments have been undertaken on mechanised, large scale, resin-crop harvesting.

The crop could be harvested wet as a silage or sugarcane type of activity, or field dried and harvested as a hay. The harvesting period is short, December and January, because it is desired to avoid leaf drop, and much of the crop will need to be stored prior to processing. The choice of harvesting system will, amongst other factors, be determined by how wet and dry crops store and how storage affects processing. We know nothing of how storage affects processing at this time. Mendel et al. (1979) argued for field drying because of a subsequent processing need to dry the Euphorbia lathyris crop, problems of fermentation of the wet crop, and increased cost of transporting the wet material. We have accepted his conclusion. Harvesting systems we have examined include small rectangular bales, stacks (3t and 1t), round bales, and loose chop.

Hitzhusen and Abdallah (1980) report relative harvesting costs (1977 US$/t) of corn stover as follows:

	3t Stack	Large round bales	Loose
Indiana	8.45	17.50	9.75
Iowa	6.65	11.30	10.44

Mendel et al. (1979) estimated harvesting and field drying costs of E. lathyris of US$(1978) 20 to 40/acre at yield of 6.8 tons(DW)/acre, i.e. $3 to $6/t(DW). We believe these costs include mowing, windrowing to field dry, chopping and loading and they seem very low.

In Australia, King and Ashby (1976) compare small square bales, round bales, 3t stacks and 1t stacks. Fig. A1 is reproduced from their work. Their costs exclude the costs of mowing and raking. At the proposed levels of planting each farm produces 1200t/yr and there is little difference between alternatives at these high levels of machinery usage.

Current costs of contract mowing, raking and baling (small rectangular bales) of lucerne hay are 80¢ to $1/bale (personal communication: R. Lobegeiger, QDPI) and at 40 bales/t costs would be $36/t left in the field.

From this variety of costs we select an estimate of $20/t on-board truck (large round bales) at any time through the year.

FIGURE A.1

Cost to bale hay (source: King and Ashby 1976)

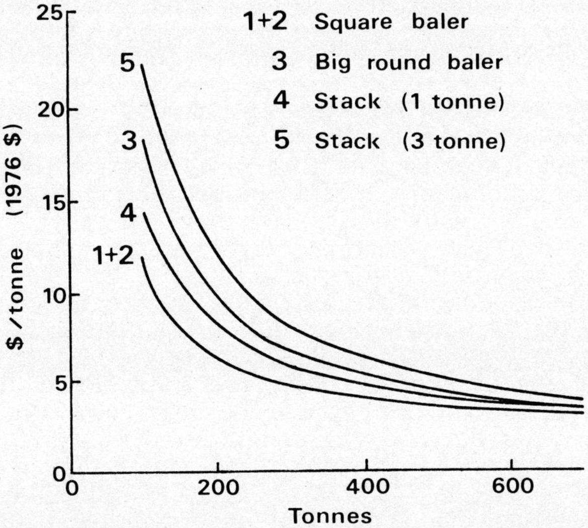

Storage Costs

The crop is assumed to be harvested over a two month period and eight months of processing plant usage needs to be stored. At a roll bale weight of 0.84t and a plant size of 300t(DW)/day storage would be needed for approximately 90 000 bales. If bale dimensions are 1.8m diameter x 1.5m length and bales are stored with lm surrounding walkways, then the total storage area is approximately 63ha. Taken over the whole area under resin crop of 30 000ha this requirement is small and no costs are charged for infield storage.

If it was necessary to store the crop under cover in pole barns to protect the sugars (and there is doubt whether they can be protected) then this could cost over $5 million of capital expenditure. This equals the capital expenditure in the resin extraction plant. It is so expensive that we do not consider it further. There might be possibilities of cheaper storage under plastic sheeting but the effects of this type of storage on the biological degradation of sugars might be high and we do not consider it further.

Transport Costs

We assume as Eckersley (1979), a farm size of 2600ha, 1200ha cropped annually and the resin crop occupying 400ha of this total cropping area. At yields of 3t(DW)/ha and a processing plant requiring 300t(DW)/d for 300 days per year, the total area cropped to resin crop is 30 000ha on 75 farms. The total farm area associated with crop is 195 000ha. If a process centre was sited in the centre of this size the circle would have a radius of 25km, the 'ideal' average haul would be 0.67 of the radius = 17km.

Siemon (1975) discusses briefly the problems of estimating average haul from the 'ideal' calculated above. In the case of a specific paper pulp mill the average haul was 4.5 times the 'ideal', and Siemon used a 'non-ideality' factor of 2.5 in his assessment of ethanol production from wood. It is not clear how much of the 'non-ideality' we have already considered in our assumptions of rotations with existing crops and the extent of cropping on the farm but this is obviously a component factor. For this study we assume a factor of 1.0 and recognize the deficiency in the analysis. Site specific studies are the only way of gathering more accurate distances.

Two sources of road transport costs are available to us, both of which allow full loading of trucks. Sugarcane haulage costs are estimated to be 15¢/t.km and the cane is hauled over distances of less than 20km in enclosed bins for 6 months of the year. Woodchips are hauled distances of between 100 and 200km at costs of 4.5¢/m^3.km (8¢/t.km) year round, again in bins with sides. There is some concern about whether round bales will be sufficiently dense to provide full loads for road transport vehicles and consequently the relevance of the costs per tonne.km that we have available. There is considerable variation in the bale density claimed for round bales but the highest densities, in excess of 220kg/m^3, compared with the minimum density necessary for full loading of 160kg/m^3. The 'effective' density of the bales on the vehicle will depend on how they are packed and how the dimensions of the bale allow efficient packing on the vehicle. Catt (1976) illustrates this point with calculations which show that, even after reducing the diameter of round bales by 20% to allow two bales to be packed lengthwise down the truck, the particular vehicle chosen only carries half of its maximum payload. We use a cost of 20¢/t.km as appropriate to the haulage of round bales (including unloading).

The preceeding points can be brought together in an equation which, for any given processing plant capacity per year, relates transport costs per tonne of feedstock to yield per hectare of feedstock. The equation is derived as follows:

p = plant feedstock throughput per year
y = yield of crop in t(DW)/ha.yr

$\therefore$ annual farm harvest area = p/y ha
c = proportion of farm under resin crop
$\therefore$ gross farm area = p/y x c ha = p/y x c x 100 km^2
$\therefore$ radius of circle emcompassing farm area = $\sqrt{p/y \times c \times 100 \times \pi}$
Ideal haul = .67 x radius
Cost of transport = \$0.20/t.km
$\therefore$ average cost of transport = $.67 \times .2\sqrt{p/100yc\pi}$

Gross Margin of Wheat Displaced

To grow resin crops on wheat land without losing money the farmer must receive the 'profit' or gross margin (gross receipts less variable costs) he made on that wheat plus the variable costs of resin crops. Table Al is an update of Eckersley (1979) using BAE indices of prices paid and received by farmers.

Table Al Gross margin of wheat displaced

Yield (t/ha)	1.1
Price ($/t)	100
Receipts ($/ha)	110
Costs ($/ha)	
Seed	4
Fertilizer	14
Spray	4
Fuel, oil, repairs and maintenance	22
	44
Gross margin	66

ACKNOWLEDGEMENTS

We wish to acknowledge the kind cooperation of many people in Australia and in USA, too numerous to list, who assisted us in discussion and by providing unpublished data. We also wish to thank Ms J.Hulme for typing and editing the report, Mrs M.Lowe and Mrs N.Geier for draughting the figures; and Ms J.Tracey and A.Hughes of the CSIRO Word Processing Centre for converting the report to camera-ready copy. We also wish to thank Mr I. Henderson of the CSIRO Science Communication Unit for the design of the cover of the report.

REFERENCES

ANGUS, J.F., CUNNINGHAM, R.B., MONCUR, M.W. and MACKENZIE, D.H. (1980) Phasic development in field crops - 1. thermal response in the seedling phase. Field Crops Res. 3:365-378.

AUSTRALIAN OILSEED CRUSHERS ASSOCIATION (1978) Oilseed production and processing. Aust. Oilseed Crushers Assoc., June.

BAILEY, D.R. and BUFFIER, B.D. (1981) Enterprise budgets for the north west of N.S.W. Complan Handbook No. 3, N.S.W. Dept. Agric., Gunnedah and Univ. New England, Armidale.

BELMARES-SARABIA, H. and JIMENEZ-VALDES, L.L. (1978) Resina de guayule: caracterizacion y possibles usos industriales. In: Guayule reencuentro en el desierto. (Consejo Nacional de Ciencia y Tecnologia, Mexico).

BENEDICT, H.M., McRARY, W.L. and SLATTERY, M.C. (1947) Response on guayule to alternative periods of low and high moisture stresses. Bot. Gaz. 108:535-540.

BENEDICT, H.M. (1949) A further study of the non-utilization of rubber as a food reserve by guayule. Bot. Gaz. 111:36-43.

BONNER, J. (1943) Effects of temperature on rubber accumulation by the guayule plant. Bot. Gaz. 105:233-243.

BUCHANAN, R.A., CULL, I.M., OTEY, F.H. and RUSSELL, C.R. (1978a) Hydrocarbon- and rubber-producing crops - evaluation of US plant species. Econ. Bot. 32(2):131-145.

BUCHANAN, R.A., CULL, I.M., OTEY, F.H. and RUSSELL, C.R. (1978b) Hydrocarbon- and rubber-producing crops - evaluation of 100 US plant species. Econ. Bot. 32(2):146-153.

BUCHANAN, R.A., OTEY, F.H., RUSSELL, C.R. and CULL, I.M. (1978c) Whole plant oils, potential new industrial raw materials. J. Am. Oil Chem. Soc. 55(9):657-662.

CALVIN, M. (1978) Hydrocarbons via photosynthesis. In: Guayule reencuentro en el desierto. (Consejo Nacional de Ciencia y Tecnologia, Mexico).

CALVIN, M. (1974) Solar energy by photosynthesis. Science 184:375-381.

CAMPOS-LOPEZ, E., NEAVEZ-CAMACHO, E. and MALDONADO-GARCIA, R. (1978) Guayule: state of knowledge. In: Guayule reencuentro ed el desierto. (Consejo Nacional de Ciencia y Tecnologia, Mexico).

CATT, W.R. (1976) Moving straw from field to factory. In: Report on straw utilisation. Conf. Oxford. January 23, 1976.

COFFEY, S.C. and HALLORAN, G.M. (1981) Euphorbia - perspectives and problems. In: National Conf. on Fuels from Crops. September 28-29, 1981. Melb.

CORNFORTH, G.C., LACEWELL, R.D., COLLINS, G.S., WHITSON, R.E. and HARDIN, D.C. (1980) Guayule - economic implications of production in the southwestern United States. Texas Agric. Exp. Stat.

CROCKER, R.L. and TRUMBLE, H.C. (1945) Investigations of guayule (Parthenium argentatum Gray) in South Australia. Bulletin 192, Council for Scientific and Industrial Research, Commonwealth of Australia.

CURTIS, O.F. (1947) Distribution of rubber and resins in guayule. Plant Physiol. 22:333-359.

DAVIDSON, B.R. (1962) Crop yields in experiments and on farms. Nature, Lond. 194:458-459.

DAVIES, L., MULLEN, C. MITCHELL, T., KEARINS, D. and SUNSTROM, B. (1980) Crop and livestock budgets, Dubbo district - 1980. Farm Business Notes No. 8, N.S.W. Dept. Agric.

DICKENSON, P.B. (1976) The application of growth modifiers to the production of latex. Outlook on Agric. 9:88-94.

DICKSON, B.T. and HARTLEY, W. (1946) Rubber growing in Australia. Trans. Inst. Rubber Ind. 22:17-24.

ECKERSLEY, P. (1979) Eastern wheatbelt farm economics. W.A. Dept. Agric.

FOSTER, K.E., McGINNIES, W.G., TAYLOR, J.G., MILLS, J.L., WILKINSON, R.R., HOPKINS, F.C., LAWLESS, E.W., MALONEY, J. and WYATT, R.C. (1980) A technology assessment of guayule rubber commercialization. Final Report prepared for the National Science Foundation, Division of Policy Research and Analysis. An OALS/MRI Report.

GOMEZ-CONTRERAS, H. (1978) Epoca y metodos adecuados para la explotacion en Mexico. In: Guayule reencuentro en el desierto. (Consejo Nacional de Ciencia y Tecnologia, Mexico).

HAMMOND, B.L. and POLHAMUS, L.G. (1965) Research on guayule (Parthenium argentatum): 1942-1959. US Dept. of Agric. Agric. Res. Serv. Tech. Bull. No. 1327.

HILLIS, W.E. and BROWN, A.G. (Eds.) (1978) Eucalypts for wood production. (CSIRO, Melbourne).

HINMAN, C.W., HOFFMAN, J.P., McLAUGHLIN, S.P. and PEOPLES, T.R. (1980) Hydrocarbon production from arid plant species. In: Proc. 1980 Ann. Meeting. (Am. Section Int. Solar Energy Soc. Inc., Newark, DE).

HITZHUSEN, F.J. and ABDALLAH, M. (1980) Economics of electrical energy from crop residue combustion with high sulphur coal. Am. J. Agric. Econ. Aug.:416-425.

JOHNSON, J.D. and HINMAN, C.W. (1980) Oils and rubber from arid land plants. Science 208:460-464.

KEMP, A.G. (1981) Brick industry - sawdust firing of kilns. Presented at: NERDDC-CSIRO R&D Workshop. Production of energy from the combustion of crop and forest residues. Canberra. July 7-8, 1981.

KING, D.J. and ASHBY, R.G. (1976) Four fodder systems - an economic appraisal. Agric. Eng. Aust. 5(1):25-28.

KOHAN, S.M. and WILHELM, D.J. (1980) Recovery of hydrocarbon-like compounds and sugars from Euphorbia lathyris. Presented at: AIChE 89th National Meeting. Portland, Oregon. August 17-20, 1980.

LEE, W., MAZIUK, J., WEEKMAN, V.W. and YURCHAK, S. (1980) Catalytic conversion of alcohols to gasoline by the Mobil process. In: Energy from biomass and wastes IV. Symp. Papers. January 21-25, 1980. Lake Buena Vista, Florida. (Inst. Gas Technology, Chicago).

LIETH, H. (1973) Primary production: terrestrial ecosystems. Human Ecology 1:303-332.

McGINNIES, W.G. (1978) Overview of guayule program. In: Guayule reencuentro en el desierto. (Consejo Nacional de Ciencia y Tecnologia, Mexico).

MENDEL, D.A., SCHOOLEY, F.A. and DICKENSON, R.L. (1979) A production cost analysis of Euphorbia lathyris, final report. (SRI International, Menlo Park).

MERRETT, A.J. and SYKES, A. (1973) The finance and analysis of capital projects. 2nd Ed. (Longman, London).

NATIONAL ACADEMY OF SCIENCES -NATIONAL RESEARCH COUNCIL. Committee on Animal Nutrition. Subcommittee on Feed Composition (1969) United States - Canadian tables of feed composition: nutrient data for United States and Canadian Feeds. (National Academy of Sciences-National Research Council, Washington, D.C.).

NATIONAL ACADEMY OF SCIENCES (1977) Guayule: an alternative source of natural rubber. (National Academy of Sciences, Washington, D.C.).

NEMETHY, E.K., OTVOS, J.W. and CALVIN, M. (1980) Hydrocarbons from Euphorbia lathyris. (Submitted to Pure Appl. Chem.).

NEMETHY, E.K., OTVOS, J.W. and CALVIN, M. (1979) Analysis of extractables from one Euphorbia. J. Am. Oil Chem. Soc. 56:957-960.

NIELSEN, P.E., NISHIMURA, H., OTVOS, J.W. and CALVIN, M. (1977) Plant crops as a source of fuel and hydrocarbon like materials. Science 198:942-944.

NISHIMURA, H., PATON, D.M. and CALVIN, M. (1980) Eucalyptus radiata as a renewable biomass. Agric. Biol. Chem. 44(10):2495-2496.

NISHIMURA, H. and CALVIN, M. (1979) Essential oil of Eucalyptus golobulus in California. J. Agric. Food Chem. 27(2):432-435.

NIVERT, J.J., GLIMP, E.M. and SYNDER, C.E. (1978) Preliminary economic analysis of guayule rubber production. In: Guayule reencuentro en el desierto. (Consejo Nacional de Ciencia y Tecnologia, Mexcio).

NIX, H.A. (1978) Availability of land for energy cropping in Australia. In: Alcohol Fuels. Conf. August 911, 1978. Sydney.

PEOPLES, T.R. and JOHNSON, J.D. (1980) Crop alternatives in semiarid regions. Presented at: Natural resources and regional development - the case of the semiarid region. Conf. Cuautla, Morelos, Mexico. October 68, 1980.

PEOPLES, T.R., ALCORN, S.M., BLOSS, H.E., CLAY, W.F., FLUG, M. HOFFMAN, J.J., LEE, C.W., LUNA, S., McLAUGHLIN, S.P., STEINBERG, M. and YOUNG, M. (1981) Euphorbia lathyris L.: A future source of extractable liquid fuel. Biosources Digest 3(3).

POLHAMUS, L.G. (1967) Plants collected and tested by Thomas A. Edison as possible sources of domestic rubber. ARS34-74, US Dept. Agric. June.

POOLE, M.L. (1980) Conservation of energy with minimum tillage. In: Energy and agriculture - the impact of changes in energy costs on the rural sector of the Australian economy. (Eds.) K.M.W. Howes and R.A. Rummery (CSIRO, Perth).

PRESCOTT, J.A. (1943) The Australian homoclime of the zone of natural occurrence of Parthenium argentatum. Trans. Roy. Soc. S.Aust. 67:312-318.

RAWLINS, W.H.M. (1981) Some economics of kenaf crop production and processing. In: Kenaf as a potential source of pulp in Australia. Proc. of Kenaf Conf. (Eds.) I.M. Wood and G.A. Stewart (CSIRO, Australia).

ROBERTS, J.D. and CASERIO, M.J. (1965) Basic principles of organic chemistry. (W.A. Benjamin Inc., New York).

ROBERTSON, G.A. (1980) Ord River Irrigation Area budgets for crop production on Kununurra clay soils. Farm Note 52 Kna 10/80. W.A. Dept. Agric.

ROSENZWEIG, M.L. (1968) Net primary production of terrestrial communities: prediction from climatic data. American Naturalist 102:67-74.

SACHS, R.M., LOW, C.B., MACDONALD, J.D., AWAD, A.R. and SULLY, M. (1981) Agronomic studies with Euphorbia lathyris - a potential source of petroleum-like products. In: Energy from biomass and wastes V. Proc. January 26-30, 1981.

SHAPTER, R.E., (1952) Rubber accumulation in guayule (Parthenium argentatum Gray) in South Australia. CSIRO Bulletin No. 270. Australia.

SIEMON, J.R. (1975) The production of solar ethanol from Australian forests. CSIRO Aust. Solar Energy Studies. Report No. 75/5.

SMALL, B.E.J. (1977) Assessing the Australian Eucalyptus oil industry. Forest & Timber 13(2):13-16.

STEWART, G.A., RAWLINS, W.H.M., QUICK, G.R., BEGG, J.E. and PEACOCK, W.J. (1981) Oilseeds as a renewable source of diesel fuel. Search 12(5): 107-115.

STEWART, G.A., GARTSIDE, G., GIFFORD, R.M., NIX, H.A., RAWLINS, W.H.M. and SIEMON, J.R. (1979a) The potential for liquid fuels from agriculture and forestry in Australia. (CSIRO, Aust.).

STEWART, G.A., GARTSIDE, G., GIFFORD, R.M., NIX, H.A., RAWLINS, W.H.M. and SIEMON, J.R. (1979b) Liquid fuel production from agriculture and forestry in Australia. Search 10(11):382-387.

TAKEDA, S. (1979) Eucalyptus oil and its implication to ignition Spark engines. Presented at: IX Int. Comm. of Agric. Eng. July 1979.

TAYLOR, K.W. (1978) Guayule processing: past and future. In: Guayule reencuentro en el desierto. (Consejo Nacional de Ciencia y Tecnologia, Mexico).

THEW, R.K. (1980) Costs and returns of summer crops in Wambo Shire - brigalow plain soil. Farm Notes No. F106/Oct 80.

TIDEMAN, J. and HAWKER, J.S. (1981a) The hydrocarbon contents of some latex-bearing plants in Australia. Search (in press).

TIDEMAN, J. and HAWKER, J.S. (1981b) In vitro propagation of latex-bearing plants. Annal. Bot. (in press).

TRAUB, H.P. (1946) Concerning the function of rubber hydrocarbon (caoutchouc) in the guayule plant, Parthenium argentatum A. Gray. Plant Physiol. 21: 425-444.

VELAZQUEZ, M.A., MARTINEZ-RAMIREZ, O. and AGUIRRE-VALDES, J. (1978) Revision historica de la produccion de hule de guayule en Mexico de 1903 a 1951. In: Guayule reencuentro en el desierto. (Consejo Nacional de Ciencia y Tecnologia, Mexico).

WAN, E.I., SIMMONS, J.A., PRICE, J.D. and NGUYON, T.D. (1979) Technical - economic assessment of the production of methanol from biomass: final research report. Vol 1 Executive Summary. Science Applications Inc., McLean, VA.

THE WEALTH OF INDIA (1950) A dictionary of Indian raw materials and industrial products. Raw Materials Vol 11. (Council of Scientific and Industrial Research, Delhi).

WEISZ, P.B., HAAG, W.O. and RODEWALD, P.G. (1979) Catalytic production of high-grade fuel (gasoline) from biomass compounds by shape-selective catalysis. Science 206:57-58.

WILLIAMS, L.R., CASALI, J., and HOME, V. (1981) Evaluation of some Australian plant species for potential as hydrocarbon producing crops. Chem. in Aust. 48(a):344-345.

YOKOYAMA, H., HAYMAN, E.P., HSU, W.J., POLING, S.M. and BAUMAN, A.J. (1977) Chemical bioinduction of rubber in guayule plant. Science 107:1076-1078.